Lecture Notes in Artificial Intelligence 8323

Subseries of Lecture Notes in Computer Science

LNAI Series Editors

Randy Goebel
University of Alberta, Edmonton, Canada
Yuzuru Tanaka
Hokkaido University, Sapporo, Japan
Wolfgang Wahlster
DFKI and Saarland University, Saarbrücken, Germany

LNAI Founding Series Editor

Joerg Siekmann
DFKI and Saarland University, Saarbrücken, Germany

T0183403

Madalina Croitoru   Sebastian Rudolph
Stefan Woltran   Christophe Gonzales (Eds.)

Graph Structures
for Knowledge Representation
and Reasoning

Third International Workshop, GKR 2013
Beijing, China, August 3, 2013
Revised Selected Papers

 Springer

Volume Editors

Madalina Croitoru
LIRMM
161, rue ADA, 34392 Montpellier Cedex 5, France
E-mail: croitoru@lirmm.fr

Sebastian Rudolph
Technische Universität Dresden
Nöthnitzer Str. 46, 01187 Dresden, Germany
E-mail: sebastian.rudolph@tu-dresden.de

Stefan Woltran
Technische Universität Wien
Favoritenstraße 9-11, 1040 Vienna, Austria
E-mail: woltran@dbai.tuwien.ac.at

Christophe Gonzales
Université Pierre et Marie Curie
4, place Jussieu, 75005 Paris, France
E-mail: christophe.gonzales@lip6.fr

ISSN 0302-9743   e-ISSN 1611-3349
ISBN 978-3-319-04533-7   e-ISBN 978-3-319-04534-4
DOI 10.1007/978-3-319-04534-4
Springer Cham Heidelberg New York Dordrecht London

Library of Congress Control Number: 2013957754

CR Subject Classification (1998): I.2, F.4, G.2, H.3

LNCS Sublibrary: SL 7 – Artificial Intelligence

Typesetting: Camera-ready by author, data conversion by Scientific Publishing Services, Chennai, India

Printed on acid-free paper

Springer is part of Springer Science+Business Media (www.springer.com)

Preface

Versatile and effective techniques for knowledge representation and reasoning (KRR) are essential for the development of successful intelligent systems. Many representatives of next generation KRR systems are based on graph-based knowledge representation formalisms and leverage graph-theoretical notions and results. The goal of the workshop series on Graph Structures for Knowledge Representation and Reasoning (GKR) is to bring together the researchers involved in the development and application of graph-based knowledge representation formalisms and reasoning techniques.

This volume contains revised selected papers of the third edition of GKR, which took place in Beijing, China on August 3, 2013. Like the previous editions, held in Pasadena, USA (2009), and in Barcelona, Spain (2011), the workshop was associated with IJCAI (the International Joint Conference on Artificial Intelligence), thus providing the perfect venue for a rich and valuable exchange.

The scientific program of this workshop included many topics related to graph-based knowledge representation and reasoning such as representations of constraint satisfaction problems, formal concept analysis, conceptual graphs, argumentation frameworks and many more. All in all, the third edition of the GKR workshop was very successful. The papers coming from diverse fields all addressed various issues for knowledge representation and reasoning and the common graph-theoretic background allowed to bridge the gap between the different communities. This made it possible for the participants to gain new insights and inspiration.

We are grateful for the support of IJCAI and we would also like to thank the Program Committee of the workshop for their hard work in reviewing papers and providing valuable guidance to the contributors. But, of course, GKR 2013 would not have been possible without the dedicated involvement of the contributing authors and participants.

November 2013

Madalina Croitoru
Sebastian Rudolph
Stefan Woltran
Christophe Gonzales

Organization

Workshop Chairs

Madalina Croitoru   Université Montpellier 2, France
Sebastian Rudolph   Technische Universität Dresden, Germany
Stefan Woltran   Technische Universität Wien, Austria
Christophe Gonzales   LIP6, Université Paris 6, France

Program Committee

Simon Andrews   Sheffield Hallam University, UK
Manuel Atencia   Inria & Université de Grenoble 1, France
Jean-François Baget   Inria & LIRMM, France
Dan Corbett   DARPA, USA
Olivier Corby   Inria, France
Cornelius Croitoru   Universitatea AI.I. Cuza, Romania
Madalina Croitoru   LIRMM, Université Montpellier II, France
Frithjof Dau   SAP, Germany
Jérôme David   Inria Rhône-Alpes, France
Paul Dunne   University of Liverpool, UK
Wolfgang Dvořák   University of Vienna, Austria
Catherine Faron Zucker   I3S, UNS-CNRS, France
Sebastien Ferre   Université de Rennes 1, France
Jerome Fortin   UMR IATE, France
Christophe Gonzales   LIP6, Université Paris 6, France
Léa Guizol   LIRMM, France
Tarik Hadzic   Cork Constraint Computation Centre, Ireland
Ollivier Haemmerlé   IRIT, Université Toulouse le Mirail, France
John Howse   University of Brighton, UK
Adil Kabbaj   INSEA, France
Mary Keeler   VivoMind, USA
Hamamache Kheddouci   Université Claude Bernard Lyon 1, France
Jérôme Lang   LAMSADE, France
Pierre Marquis   CRIL-CNRS & Université Artois, France
Tomasz Michalak   University of Warsaw, Poland
Nir Oren   University of Aberdeen, UK
Bruno Paiva Lima Da Silva   LIRMM, France
Simon Polovina   Sheffield Hallam University, UK
Julien Rabatel   GREY, Université de Caen, France
Sebastian Rudolph   Technische Universität Dresden, Germany
Eric Salvat   IMERIR, France

Gem Stapleton	University of Brighton, UK
Nouredine Tamani	UMR IATE INRA-SUPAGRO, France
Dan Tecuci	Siemens Corporate Research, USA
Michaël Thomazo	Technische Universität Dresden, Germany
Thanh Tran	San Jose State University, USA
Srdjan Vesic	CRIL-CRNS, France
Nic Wilson	University College Cork, Ireland
Stefan Woltran	Technische Universität Wien, Austria
Pierre-Henri Wuillemin	LIP6, France

Additional Reviewers

Mustafa Al-Bakri
Peter Chapman

Table of Contents

Implementation of a Knowledge Representation and Reasoning Tool Using Default Rules for a Decision Support System in Agronomy Applications

Patrice Buche[1], Virginie Cucheval[2], Awa Diattara[3],
Jérôme Fortin[4], and Alain Gutierrez[5]

[1] INRA IATE
LIRMM GraphIK, France
Patrice.Buche@supagro.inra.fr
[2] CTFC Poligny, France
v-cucheval@ctfc.fr
[3] INRA IATE, France
diattaraawa@gmail.com
[4] Université Montpellier II
IATE/LIRMM GraphIK, France
fortin@polytech.univ-montp2.fr
[5] CNRS LIRMM, France
alain.gutierrez@lirmm.fr

Abstract. This is an application paper in which we propose to use an extended version of the conceptual graph framework to represent and reason on expert knowledge for cheese making. In this extension, we propose to use default rules to represent specific pieces of knowledge. We use the CoGui software to manage conceptual graphs in the application from the CTFC data expressed in Freeplan. A specific end user interface has been designed on top of CoGui to ensure that end-user experts in cheese making can use it in a convenient way.

1 Introduction

The CTFC (Centre Technique des Fromages Comtois, which means Technical Centre for Comtois Cheese) is a technical centre in the east part of France that helps traditional cheese companies to manage cheese making process. For example when a cheese maker finds a problem in some cheeses (as bitterness or a consistency default), some diagnostic can be made by the CTFC in order to propose a solution to the cheese maker. For this purpose, the CTFC needs to capitalize knowledge and experience about traditional cheese making process. This knowledge is owned, mainly in an informal way, by a group of experts employed by CTFC. Capitalizing this knowledge is a crucial topic to maintain the quality of specific cheese making process as Comté, Morbier, Bleu de Gex or Mont d'or. It is also important to capitalize expert knowledge because some processes are not fashionable at a given moment but may become fashionable in the future. For example the traditional Comté had some holes like Swiss Gruyère around

M. Croitoru et al. (Eds.): GKR 2013, LNAI 8323, pp. 1–12, 2014.

15 years ago. There is no more hole nowadays in Comté, so if the knowledge on the way to obtain Comté with holes is not capitalized, nobody will be able to do it again in the future. As this knowledge is not always formally represented, the risk to loose this information is important.

This is why the CTFC has decided to develop a knowledge management system that allows to capitalize technological and scientific knowledge about cheese making. The implementation of this system requires a methodology to collect the operational knowledge of technical experts and identify the underlying scientific knowledge. A tool to structure this knowledge and a tool able to reason on this knowledge to provide answers to cheese experts queries are also required.

A methodological approach has been defined to collect the expert knowledges through two kinds of interviews: on the one hand, individual interviews and on the other hand, collective and contradictory ones. This approach is now operational and the expert knowledge resulting from those interviews is represented in a tree structure using a free mind mapping software called Freeplan. Once the knowledge collected and entered in Freeplan, a "scientific validation" session is done in collaboration with INRA researchers (National Institute of Agronomy) of the URTAL unit in Poligny. A lack of Freeplan is that pieces of information must be stored in separate files as the information is voluminous and no reasoning tool is available to use this information. For example, informations may be easily displayed in Freeplan mind map. But, if the user wants to find the list of possible corrective actions in order to control a given property of the cheese (quality or default), it becomes difficult to display all the mind maps in a short time. Therefore, automatic operations are required to analyze the knowledge. As the CTFC wants to manage the knowledge associated with around 50 properties of cheese, a complementary tool that permits to query automatically the information is required.

To reach this aim, our approach consists in translating the knowledge described in Freeplan mind maps into the conceptual graph formalism which permits to perform automatic reasoning. To implement the tool, we use CoGui free software, which is a conceptual graph editor that permits to manage the terminology of an application domain, the facts, the rules and more recently default rules. We developed a user interface on top of CoGui to ensure that end-users of the application can easily use it without knowing anything about the conceptual graph formalism.

The paper is organised as follows : the next section presents how the CTFC actually structures its knowledge using Freeplan mind maps. In Section 3, we recall the conceptual graph formalism and explain how the expert knowledge of the CTFC can be entered in CoGui using rules and default rules. Section 4 presents the end-user application, built on top of CoGui, which permits experts to access to the knowledge without particular background formalism. We conclude the paper in the last section.

2 Structuring CFTC Knowledge Using Freeplane

Currently, expert and scientific CTFC knowledge are structured in a tree using the open software Freeplane. The structure of the tree is performed in such a way that from a given descriptor (defined as a desired or not desired property of cheese), we list all explanatory mechanisms, from the most general to the most specific and leading finally to some corrective actions (see Figure 1). We read this figure as follows : The descriptor (which may be, for example, a default in the cheese making process) can be explain by Explanatory Mechanism 1 or by Explanatory Mechanisms 2... Each mechanism can be explained by several sub-mechanisms. This way to consider the different mechanisms to explain a default naturally leads to a tree structure. Finally a mechanism is associated with a particular corrective action, which can permit to control the descriptor (for example, to avoid a default if the descriptor represents a default) which is the root of the tree.

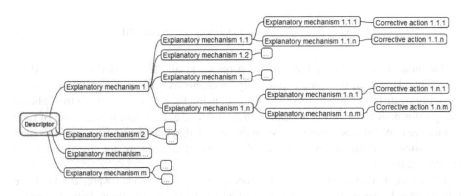

Fig. 1. Tree structure of CTFC knowledge expressed using the FreePlan software

Moreover, relationships between explanatory mechanisms and more complex actions must be taken into account. We can cite for example the joint effects or the compensatory actions, which are presented in the following.

We're talking about joint effect when two or more explanatory mechanisms should be grouped together to affect the descriptor or the $n - 1$ mechanism level. The effect is expressed in the Freeplan tree by the creation of the relationship "AND". An example of joint effect in given in Figure 2, which represents that the descriptor "salt intake after 15 days is low" is explain by the mechanism "low salt absorption", which is explained by "low granularity of cheese crust", which is jointly explained by "using smooth mold" and "high duration contact between cheese and mold ($> 8h$)".

A corrective action is defined as a way to control a descriptor (for example to correct a default) for the next manufacturing. On another hand, a compensatory action is an action that is taken to control a descriptor (for example to correct a default) during

Fig. 2. Representation of a joint effect

the current manufacturing. Compensatory actions are expressed on the tree using the relationship "UNLESS" ("SAUF SI" in french). Figure 3 presents an example of compensatory action. It means that we can explain that a "badly realized brining" is due to a "use of a low concentration of brine ($< 250g/L$))" unless the "brining time extended".

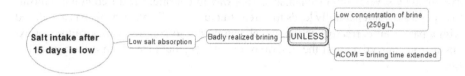

Fig. 3. Example of a compensatory action (ACOM)

The mind map model of Freeplane can represent all the knowledge expressed by the CTFC. However, the same explanatory mechanisms may appear in different trees (concerning different descriptors). In this case, using Freeplan model leads to duplicate many informations, which is not satisfactory for several reasons : it is a waste of time to manage duplications, especially when the knowledge must be updated, and this may lead to some inconsistencies of the knowledge base if updates are not propagated in all the duplicates.

Moreover, no reasoning can be performed using the Freeplan software. To overcome this drawback, we propose to use the CoGui software which allows both the representation of knowledge and reasoning with the conceptual graph model. It includes all the required elements to represent the explanation mechanisms and the joint effect relationships.

3   Structuring Knowledge of CTFC with CoGui

In general, a knowledge representation formalism should satisfy three essential criteria. First, it must allow to represent the knowledge of an application domain in a declarative way, meaning that the semantics associated with the knowledge must be defined regardless of the programs that use the knowledge base. The second criterion is that for reasoning, it must allow to make inferences that are based on logic. The third criterion is to structure the knowledge in an unambiguous way: it means that the informations linked in a semantic way should be grouped and that the knowledge of different natures must be clearly differentiated. Conceptual graphs (noted CG in the following) and CoGui (one CGs editor) satisfy all these three essential criteria.

3.1   The Conceptual Graph Formalism

The conceptual graph formalism Sowa, 1984, Chein and Mugnier, 2009 is a knowledge representation and reasoning formalism based on labelled graphs. In its simplest form, a CG knowledge base is composed of two parts: the *support*, which encodes terminological knowledge –and constitutes a part of the represented domain ontology– and *basic conceptual graphs* built on this support, which express assertional knowledge, or *facts*. The knowledge base can be further enriched by other kinds of knowledge built on the support: in this paper, we will consider two kinds of *rules*: "usual rules" and CG defaults, which lead to non-monotonic reasoning.

The support. It provides the ground vocabulary used to build the knowledge base. It is composed of a set of *concept types*, denoted by T_C, and a set of *relation types* (or simply *relations*), denoted by T_R. Relation types represent the possible relationships between concept instances, or properties of concept instances for unary relation types. T_C is partially ordered by a *kind of* relation, with $\top$ being its greatest element. T_R is also partially ordered by a *kind of* relation, with any two comparable relation types having necessarily the same arity (i.e., number of arguments). Each relation type has a signature that specifies its arity and the maximal concept type of each of its arguments.

CoGui (Conceptual Graphs Graphical User Interface)[1] is a software which permits to build knowledge bases as CGs. It provides a Java graphical interface for editing support, CGs , rules and constraints. The knowledge base can be serialized in a XML predefined format called CogXML. It includes a querying system based on forward chaining mechanism for querying a knowledge base.

Definition of the support for the CTFC application. We recall that all the knowledge of CTFC is represented as trees in Freeplane. To model this knowledge in CoGui, we first define the basic vocabulary as concept types and relation types. The concept types used to represent the specific vocabulary of the CTFC application are of three categories:

- **cheese descriptors** with two subtypes of concepts: the sensory descriptors and analytical descriptors,
- **explanatory mechanisms** including three subtypes of concepts: the states of milk, the states of the cheese and the process parameters,
- **actions** with two subtypes of concepts:compensatory actions which appear in UNLESS relationships and corrective actions associated with the last level of explanatory mechanisms in Freeplane trees.

We have identified two types of relations:

- **unary relations** with several subtypes to qualify the set of different sensory and analytical descriptors and corrective action mechanisms
- **binary relations** with several subtypes including: the relationship *"is explained by"* that connects a descriptor or an explanatory mechanism to an explanatory mechanism, the relationship *has for corrective action* that connects an explanatory mechanism of last level in the Freeplan tree to a **corrective action**.

Figure 4 shows on its left-side the hierarchy of concept types and on its right-side the hierarchy of relations. The current version of the CTFC knowledge base is composed of 114 concept types and 39 relation types.

---

[1] http://www2.lirmm.fr/cogui/

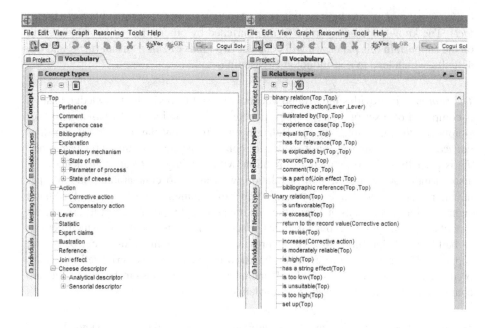

Fig. 4. Hierarchy of concept types and relation types for the CTFC application

Basic conceptual graphs. A basic CG is a bipartite graph composed of:

(i) a set of *concept nodes* (pictured as rectangles), which represent entities, attributes, states or events;
(ii) a set of *relation nodes* (pictured as ovals), which express the nature of relationships between concept nodes;
(iii) a set of *edges* linking relation nodes to concept nodes;
(iv) a *labelling* function, which labels each node or edge: the label of a concept node is a pair $t : m$, where t is a concept type and m is a marker; the label of a relation node is a relation type; the label of an edge is its rank in the total order on the arguments of the incident relation node.

Furthermore, a basic CG has to satisfy relation signatures: the number of edges incident to a relation node is equal to the arity of its type r, and the concept type assigned to its neighbour by an edge labelled i is less or equal to the ith element of the signature of r. The marker of a concept node can be either an identifier referring to a specific individual or the generic marker (denoted $*$) referring to an unspecified instance of the associated concept type. The generic marker followed by a variable name (for instance $*x$) is used in a basic CG or a rule to indicate that the instance (noted x) represented by several concept nodes is the same. A basic CG without occurrence of the generic marker is said to be totally instantiated.

Logical translation. Conceptual graphs have a logical translation in first-order logic, which is given by a mapping classically denoted by ϕ. ϕ assigns a formula $\phi(S)$ to a support S, and a formula $\phi(G)$ to any basic CG G on this support. First, each concept or relation type is translated into a predicate (a unary predicate for a concept type, and a predicate with the same arity for a relation type) and each individual marker occurring on the graphs is translated into a constant. Then, the *kind of* relation between types of the support is translated by logical implications.

Given a basic conceptual graph G on S, $\phi(G)$ is built as follows. A distinct variable is assigned to each concept node with a generic marker. An atom of the form $t(e)$ is assigned to each concept node with label $t : m$, where e is the variable assigned to this node if $m = *$, otherwise $e = m$. An atom of the form $r(e_1, \ldots, e_k)$ is assigned to each relation node with label r, where e_i is the variable or the constant corresponding to the ith neighbour of the relation. $\phi(G)$ is then the existential closure of the conjunction of all atoms assigned to its nodes.

Specialization relation, homomorphism. Any set of conceptual graphs is partially pre-ordered by a *specialization relation*, which can be computed by a *graph homomorphism* (allowing the restriction of the node labels), also called *projection* in the conceptual graph community. The specialization relation, and thus homomorphism, between two graphs, corresponds to the logical entailment between the corresponding formulas, i.e., there is a homomorphism from G to H both built on a support S if and only if $\phi(G)$ is entailed by $\phi(H)$ and $\phi(S)$ (see e.g., Chein and Mugnier, 2009 for details)[2].

The specialization relation is particularly interesting in our application to query the knowledge base. For example, an expert will be able to look for any kind of explanatory mechanisms or can specify that he/she wants only to retrieve the explanatory mechanisms associated with specific parameters of the process.

Basic CG rules. Basic CG rules Salvat and Mugnier, 1996 are an extension of basic CGs. A CG rule (notation: $R = (H, C)$) is of the form "if H then C", where H and C are two basic CG (respectively called the *hypothesis* and the *conclusion* of the rule), which may share some concept nodes. Generic markers referenced by a variable name as $*x$ refer to the same individual in the hypothesis and in the conclusion. Graphically, it can be represented by a single bicolored basic CG.

A rule R is applicable to a basic CG G if there is a homomorphism from its hypothesis to G. Let π be such a homomorphism. Then, the *application of R on G according to π* produces a basic CG obtained from G by adding the conclusion of R according to π, i.e., merging each frontier node c of the added conclusion with the node of G that is image of c by the homomorphism π.

The mapping ϕ to first-order logic is extended to CG rules. Let $R = (H, C)$ be a CG rule, and let $\phi'(R)$ denote the conjunction of atoms associated with the basic CG underlying R (all variables are kept free). Then, $\phi(R) = \forall x_1 \ldots \forall x_k (\phi'(H) \rightarrow (\exists y_1 \ldots \exists y_q \phi'(C)))$, where $\phi'(H)$ and $\phi'(C)$ are the restrictions of $\phi'(R)$ to the nodes of H and C respectively, $x_1, \ldots, x_k$ are the variables appearing in $\phi(H)$ and $y_1, \ldots, y_q$ are the variables appearing in $\phi(C)$ but not in $\phi(H)$.

Given a set of rules $\mathcal{R}$, basic CGs G and H (representing for instance a query and a set of facts), all defined on a support S, $\phi(G)$ is entailed by $\phi(H)$, $\phi(S)$ and the logical

---

[2] Note that, for the homomorphism completeness part, H has to be in normal form: each *individual* marker appears at most once in it, i.e., there are no two concept nodes in H representing the same identified individual.

formulas assigned to $\mathcal{R}$ if and only if there is a sequence of rule applications with rules of $\mathcal{R}$ leading from H to a basic CG H' such that there is a homomorphism from G to H' (in other words, by applying rules to H, it is possible to obtain H' which entails G).

When a rule is applied, it may create new individuals (one for each generic concept node in its conclusion, i.e., one for each existential variable y_i in the logical translation of the rule). In the following, we will assume that all facts (represented as basic CGs) are completely instantiated. Then, when a rule is applied, we will instantiate each new generic concept node created, by replacing its generic marker with a new individual marker (which can be seen as a Skolem function, moreover without variable in this case). This way of doing will allow us to represent CG defaults in a simpler way (see the next section).

Translation of Freeplane trees into CG rules. In the CTFC application, each Freeplane tree is translated into a set of CG rules. To do that, each elementary explanatory mechanism (defined as a couple of explanatory mechanisms linked by the *"is explained by"* relationship) will be translated into a new CG rule. By this way, a given information about an explanatory mechanism has to be entered only once, even if is it used to explain several descriptors. Moreover as we will see in the next sections, it will be possible to reconstruct easily all the CTFC knowledge trees, just by defining a descriptor (root of the tree) as a graph fact and by applying all the rules to it.

Figure 5 is an example of a rule meaning that "a low salt rate in ripened cheese" is explained by "a low salt intake during ripening". This CG rule is associated with one of the explanatory mechanisms represented in the Freeplan knowledge tree partially given in Figure 3. Note that the hypothesis and the conclusion of the rule are presented as two different CGs, linked together by a co-reference link (represented by a dashed line in the figure 5).

Default Rules in the Conceptual Graph Formalism. We now present an extension of CG rules, which has been introduced in Baget et al., 2009, Baget and Fortin, 2010 and allows for default reasoning. It can be seen as a graphical implementation of a subset of Reiter's default logic Reiter, 1980: indeed, we restrict the kind of formulae that can be used in the three components of a default. These three components are called the hypothesis H, the conclusion C and some justifications $J_1, \cdots, J_k$. We can deal directly with non–closed defaults, i.e., without instantiating free variables before processing the defaults. In Reiter's logic, the application of a default is subject to a consistency check with respect to current knowledge: each justification J has to be consistent with the current knowledge, i.e., $\neg J$ should not be entailed by it. In CG defaults, justifications are replaced by graphs called *constraints*; a constraint C can be seen as the negation of a justification: C should not be entailed by current knowledge.

Definition 1 (CG defaults). *A CG default is a tuple $D = (H, C, C_1, \ldots, C_k)$ where H is called the hypothesis, C the conclusion and $C_1, \ldots,$ and C_k are called the constraints of the default; all components of D are themselves basic CGs and may share some concept nodes.*

Briefly said, H, C and each C_i respectively correspond to the prerequisite, the consequent and the negation of a justification in a Reiter's default. H, C and the C_i's can

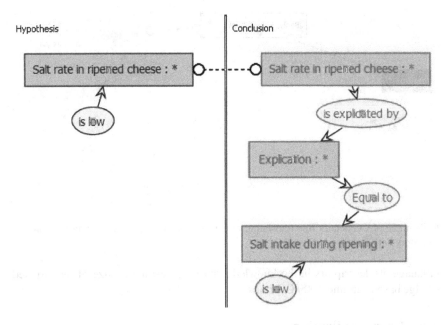

Fig. 5. Example of a standard CG rule associated with an elementary explanatory mechanism

share some concept nodes that have the same marker. These markers can be individual or generic, in which case the identification of the concept nodes is made by comparing the name of the variable associated with this generic marker.

The intuitive meaning of a CG default is rather simple: "for all individuals $x_1 \ldots x_k$, if $H[x_1 \ldots x_k]$ holds true, then $C[x_1 \ldots x_k]$ can be inferred provided that no $C_i[x_1 \ldots x_k]$ holds true". If we can map by homomorphism the hypothesis H of a default to a fact graph G, then we can add the conclusion of the default according to this homomorphism (as in a standard rule application), unless this homomorphism can be extended to map one of the constraints from the default. As already pointed out, while the negation of a justification in a Reiter's default should not be entailed, in a CG default the constraint itself should not be entailed.

The entailment mechanism is based on the construction of a default derivation tree, we let the reader refer to Baget et al., 2009, Baget and Fortin, 2010 for more precise details about default entailment mechanism.

Representing compensatory actions in the CTFC application using CG default rules. In the CTFC application, the default rules permit to model the compensatory actions in the CG knowledge base. Figure 6 shows how to represent a default rule in CoGui. This CG default rule is the translation in the CG model of the Freeplan knowledge tree presented in Figure 3. The hypothesis and the conclusion of the default are shown as for standard rule, while the justification is put in grey (left part of the rule).

The current version of the CTFC knowledge base is composed of 67 CG rules including 3 CG default rules associated with 3 descriptors. As the objectives of the CTFC

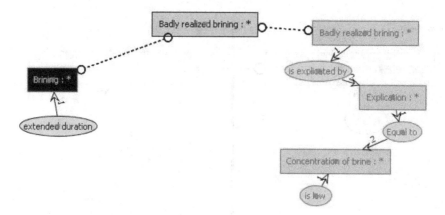

Fig. 6. Example of CG default rule translating in the CG model the Freeplan representation of a compensatory action shown in Figure 3

is to manage 50 descriptors in the knowledge base, the estimated size of the targeted knowledge base is around 1000 CG rules.

4   End-User Application

The objective of this section is to describe the implementation an application for the CTFC technicians that have no background on knowledge representation models and especially on the CG model. Therefore, an end-user application must be proposed on top of the CoGui software used to manage the CG knowledge base.

Functional requirements address the specifications defined for the CTFC application. This application must be as transparent as possible regarding the CG knowledge formalism. The information should be presented to the end-user in a similar way as to the knowledge tree presentation in Freeplan.

For our system, we have identified the following needs:

- **Display of the list of possible corrective actions associated with a given descriptor:** an expert should be able to choose a cheese descriptor from a provided list, and then get all the corrective actions associated with it. For example, if in the presence of the descriptor corresponding to the cheese default "salt intake after 15 days is low", the expert wants to know the different corrective actions that can be used to solve the problem. Moreover, for a given corrective action, the expert wants to know which compensatory actions would be avoided if the corrective action is implemented.
- **Visualization of the path from a descriptor to the associated corrective actions:** for a given descriptor, the user must have access to the "*path*" that links the descriptor to the associated corrective actions in order to visualize the various intermediate explanatory mechanisms.
- **Impact of the implementation of a corrective action:** the expert wants to know the different descriptors that may be impacted by the choice of a particular corrective action.

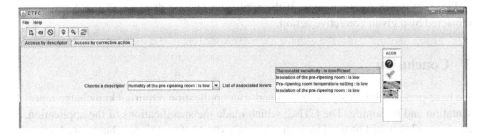

Fig. 7. Screen-shot of the final application showing how the corrective actions (4 actions in this example) of a given descriptor are presented

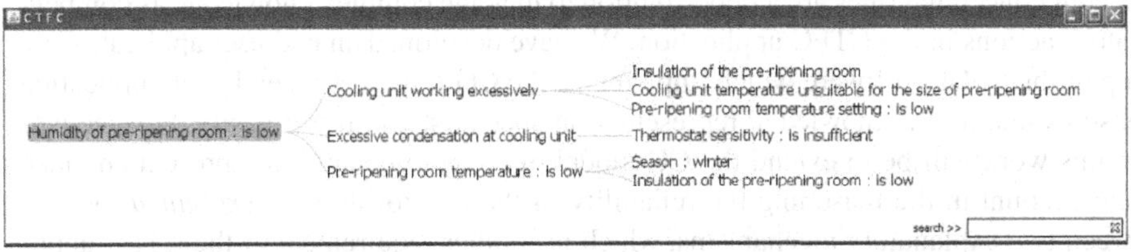

Fig. 8. Screen-shot of the final application showing a particular path in a knowledge tree

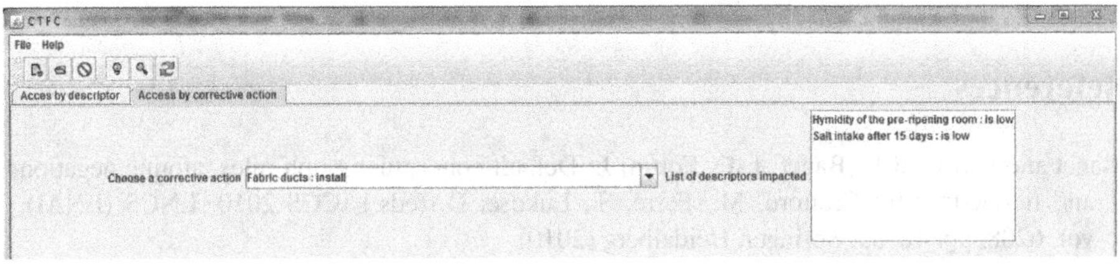

Fig. 9. Screen-shot of the final application showing all the descriptors impacted by a given corrective action

We obviously see that these requirements are all quite easy to fulfil when we represent the knowledge as CG rules with CoGui but was impossible to take into account when the knowledge model by CTFC was represented using the mind map model of Freeplan.

An end-user application has been developed in order to fulfil to all these specifications. It permits to demonstrate that CoGui can be used as an internal knowledge base engine in a dedicated application software and there is no need for the final user to be familiar neither with CoGui nor with the conceptual graph formalism. Figure 7 and 8 shows 2 screen copy of the final end-user application. In this last figure, we see that we can reconstruct the tree structure mind map model of Freeplan from the CG rules defined in CoGui. Doing that is quite easy as it only requires to define the root of the tree (the descriptor) as a simple conceptual graph, and to saturate the knowledge base composed of CG rules to construct the tree.

Figure 9 shows a screen copy of the application which displays all the descriptors impacted by a given corrective action.

5 Conclusion

We have presented in this paper a real end-user application requiring knowledge representation and reasoning. The CTFC, which made the specifications of the application, needs a tool that will be used by technological experts of traditional cheese production. We showed that on the one hand the application has a strong formal background based on conceptual graphs formalism and on the other hand that this formalism is relevant to model a complex knowledge information system. Especially, we have shown that the default conceptual rules are a good solution to manage complex knowledge as compensatory actions in the CTFC application. We have developed an end user application on top of the CoGui software which implements the CG model. This end-user application ensures that a non knowledge representation specialist can use it easily. Perspectives of this work will be to extend the CG model in order to be able to represent and take into account in the reasoning the reliability of the relationship *"is explained by"* between two explanatory mechanisms, which is another requirement of the CTFC application. Finally, another perspective of this work, suggested by the CTFC application, is to develop semi-automatic translation tools of semi-structure knowledge representation models, as Freeplan mind maps, into formal representation models as the CG model.

References

[Baget and Fortin, 2010]Baget, J.-F., Fortin, J.: Default conceptual graph rules, atomic negation and tic-tac-toe. In: Croitoru, M., Ferré, S., Lukose, D. (eds.) ICCS 2010. LNCS (LNAI), vol. 6208, pp. 42–55. Springer, Heidelberg (2010)

[Baget et al., 2009]Baget, J.-F., Croitoru, M., Fortin, J., Thomopoulos, R.: Default conceptual graph rules: Preliminary results for an agronomy application. In: Rudolph, S., Dau, F., Kuznetsov, S.O. (eds.) IÇCS 2009. LNCS (LNAI), vol. 5662, pp. 86–99. Springer, Heidelberg (2009)

[Chein and Mugnier, 2009]Chein, M., Mugnier, M.-L.: Graph-based Knowledge Representation and Reasoning. Computational Foundations of Conceptual Graphs. Advanced Information and Knowledge Processing Series. Springer, London (2009)

[Reiter, 1980]Reiter, R.: A logic for default reasoning. Artificial Intelligence 13, 81–132 (1980)

[Salvat and Mugnier, 1996]Salvat, E., Mugnier, M.-L.: Sound and complete forward and backward chaining of graph rules. In: Eklund, P., Mann, G.A., Ellis, G. (eds.) ICCS 1996. LNCS, vol. 1115, pp. 248–262. Springer, Heidelberg (1996)

[Sowa, 1984]Sowa, J.F.: Conceptual Structures: Information Proc. in Mind and Machine. Addison–Wesley (1984)

Information Revelation Strategies
in Abstract Argument Frameworks
Using Graph Based Reasoning

Madalina Croitoru[1] and Nir Oren[2]

[1] University of Montpellier 2, France
croitoru@lirmm.fr
[2] University of Aberdeen, UK
n.oren@abdn.ac.uk

Abstract. The exchange of arguments between agents can enable the achievement of otherwise impossible goals, for example through persuading others to act in a certain way. In such a situation, the persuading argument can be seen to have a positive utility. However, arguments can also have a negative utility — uttering the argument could reveal sensitive information, or prevent the information from being used as a bargaining chip in the future. Previous work on arguing with confidential information suggested that a simple tree based search be used to identify which arguments an agent should utter in order to maximise their utility. In this paper, we analyse the problem of which arguments an agent should reveal in more detail. Our framework is constructed on top of a bipolar argument structure, from which we instantiate *bonds* — subsets of arguments that lead to some specific conclusions. While the general problem of identifying the maximal utility arguments is NP-complete, we give a polynomial time algorithm for identifying the maximum utility bond in situations where bond utilities are additive.

1 Introduction

When participating in dialogue, agents exchange arguments in order to achieve some goals (such as convincing others of some fact, obtaining a good price in negotiation, or the like). A core question that arises is what arguments an agent should utter in order to achieve these goals. This dialogue planning problem is, in most cases, computationally challenging, and work on argument strategy [2,6,7,9] has identified heuristics which are used to guide an agent's utterances.

In this paper we consider a scenario where an agent must select some set of arguments to advance while taking into account the cost, or benefit, associated with revealing the arguments. [7] deals with a similar situation, and give the example of a government attempting to convince the public that weapons of mass distraction exist in some country. They assume that doing so will result in a positive utility gain. In order to back up their claims, the government must give some further evidence, and have a choice of arguments they can advance in doing so, ranging from citing claims made by intelligence resources on the ground, to

M. Croitoru et al. (Eds.): GKR 2013, LNAI 8323, pp. 13–20, 2014.
© Springer International Publishing Switzerland 2014

showing spy satellite photographs, to withdrawing the claims. Each of these arguments has an associated utility cost, and the government must therefore identify the set of arguments which will maximise its utility. In such a situation, it is clear that advancing all arguments is not always utility maximising for an agent, and [7] utilise a one-step lookahead heuristic to maximise utility while limiting computational overhead.

Unlike [7], in this paper we assume that an agent can advance several arguments simultaneously within a dialogue, and must justify these arguments when advancing them. We therefore seek to identify all arguments that an agent must advance at a specific point in time. To do so, we utilise a bipolar argument framework [3] to allow us to deal with both attacks between arguments and argument support.

We solve our problem through a translation of the argument structure into a graph structure, and then utilise graph operations in order to calculate the appropriate set of arguments to advance. Such a translation also allows us to derive an interesting result with regards to the complexity of bonds calculation.

In the next section we introduce the concept of a *bond* — a set of arguments that should be introduced together by an agent. We then examine the problem of computing a maximum utility bond. The paper concludes with a discussion of possible extensions.

2   Bonds

An argument can have several possible justifications. In the context of a dialogue, it is clearly desirable to advance the *maximal utility* justification. Importantly, this justification often does not coincide with the maximal justification in the set theoretic sense, as the utility of the entire set of arguments might be smaller than the utility of a subset of these arguments. A *bond* is then precisely the maximal utility justification for an argument. This is illustrated by the following informal example.

Example 1. A student is asked to justify why they did not hand in their homework on time, and can respond in several ways. First, they could claim they had done the homework, but that their new puppy ate it. Second, they could explain that they were ill. Furthermore, they could blame this illness on either a nasty virus they had picked up, or due to a hangover caused by over exuberance during the weekend. Clearly, providing all these reasons will not engender as much sympathy as simply blaming the virus. The latter therefore forms a maximal utility justification aimed at obtaining the teacher's sympathy, and forms a bond.

Bonds originate through the possibility that multiple lines of argument yield the same result, and that some of these have different utility costs and benefits when compared to others. A bond is made up of the subset of paths that maximise the agent's utility.

We situate bonds within Bipolar argumentation frameworks [1]. Unlike standard Dung argumentation frameworks, bipolar frameworks explicitly consider

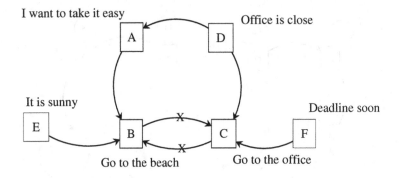

Fig. 1. Bipolar Argumentation System with argument valuations

both support and attack between arguments. We begin by formalising Bipolar frameworks, following which we introduce the notion of a *coalition* [3]. Such coalitions can be thought of as the set of all justifications for an argument. Bonds are then a subset of justifications from within a coalition, representing a single line of justifications to the conclusion.

Definition 1. *(Bipolar Argument Framework) An abstract bipolar argument framework is a tuple BAF = $(\mathcal{A}, \mathcal{R}_{def}, \mathcal{R}_{sup})$ where $\mathcal{A}$ is a set of arguments; $\mathcal{R}_{def}$ is a binary relation $\subseteq \mathcal{A} \times \mathcal{A}$ called the defeat relation; $\mathcal{R}_{sup}$ is a binary relation $\subseteq \mathcal{A} \times \mathcal{A}$ called the support relation. A bipolar argument framework obeys the constraint that $\mathcal{R}_{def} \cap \mathcal{R}_{sup} = \emptyset$.*

Definition 2. *(Coalitions) Given a bipolar argument framework BAF = $(\mathcal{A}, \mathcal{R}_{def}, \mathcal{R}_{sup})$, a coalition is a set of arguments $\mathcal{C} \subseteq \mathcal{A}$ such that all of the following conditions hold.*

1. *The subgraph $(\mathcal{C}, \mathcal{R}_{sup} \cap \mathcal{C} \times \mathcal{C})$ is connected.*
2. *$\mathcal{C}$ is conflict free.*
3. *$\mathcal{C}$ is maximal with respect to set inclusion.*

Definition 3. *(Bonds) Given a bipolar argument framework BAF = $(\mathcal{A}, \mathcal{R}_{def}, \mathcal{R}_{sup})$, and $\mathcal{C} \subseteq \mathcal{A}$, a coalition within BAF, a subset $\mathcal{B} \subseteq \mathcal{C}$ is a bond if and only if there is no $a \in \mathcal{C}$ such that for some $b \in \mathcal{B}$, $(b, a) \in \mathcal{R}_{sup}$.*

Example 2. To illustrate these concepts, consider the bipolar argument framework illustrated in Figure 1. Here, arrows with crosses indicate attacks between arguments, while undecorated arrows represent support. Let us also associate utilities with each argument in the system as follows: $u(A) = 11$, $u(B) = 5$, $u(C) = -2$, $u(D) = -8$, $u(E) = 4$ and $u(F) = -10$.

Figure 2 depicts the two coalitions found in this framework, namely $\mathcal{C}_1 = \{D, A, B, E\}$ and $\mathcal{C}_2 = \{A, D, C, F\}$. The utility associated with the latter is 9, and with the former, 12. Now consider the following bond (which is a subset of $\mathcal{C}_1$): $\{A, E, B\}$. Its utility is 20, and in a dialogue, these are the arguments that an agent should advance.

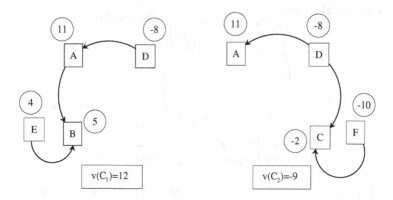

Fig. 2. Coalitions in Argumentation System in Figure 1

With the definition of bonds in hand, we now turn our attention to how the maximum utility bond — $\{A, E, B\}$ in the previous example — can be computed.

3   Identifying Bonds

A naïve approach to computing maximal utility begins with a coalition C and enumerating its bonds, beginning with arguments which do not support other arguments (these are bonds of cardinality 1), then considering bonds with a single support (i.e. bonds of cardinality 2), and so on. Once all bonds are computed, the maximal utility ones are identified and returned. Clearly, this approach is, in the worst case, exponential in the number of arguments in the domain.

We can construct a polynomial time solution by treating the problem as a maximum flow problem on an appropriate network. The complexity of this type of algorithm is $O(|C|^3)$, where $|C|$ is the number of nodes in C if we apply a push-relabel algorithm [5]. We begin by considering a *induced support graph* by the coalition over the original graph, defined next. The graphs of Figure 2 are examples of such induced support graphs.

Definition 4. *(Induced Support Graph) Let $BAF = (\mathcal{A}, \mathcal{R}_{def}, \mathcal{R}_{sup})$ be an abstract bipolar argumentation framework and $C \subseteq \mathcal{A}$ a coalition in BAF. We define the graph G_C^{BAF} (the induced support graph by C) as the graph $G_C^{BAF} = (N_C, E_{sup}|_C)$ where:*

- *Each node in N_C corresponds to an argument in the coalition C and*
- *The edges are only the support edges restricted to the nodes in C (denoted by $E_{sup}|_C$).*

Within such an induced support graph, a bond is a set $N_\mathcal{B} \subseteq N_C$, where for each $n \in N_\mathcal{B}$, and for each edge $(n, m) \in E_{sup}|_C$, it is also the case that $m \in N_\mathcal{B}$. Since we always compute the induced support graph with respect to some underlying bipolar argument framework, we will denote G_{BAF}^C as G^C.

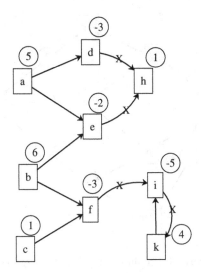

Fig. 3. Bipolar Argumentation System and its valuation

Additionally, we denote the utility of an argument corresponding to a node n in the graph as $u(n)$. The utility of a set B of arguments is defined as $u(N_B) = \sum_{n \in N_B} u(n)$. For convenience, we denote those nodes associated with a positive utility by N_C^+, and those with a negative utility by N_C^-.

We now show how the problem of finding the maximum utility bond of a coalition can be solved by reducing it to a minimum-cut computation on an extended network $G_{extended}^{\mathcal{C}}$. The idea is to construct this new network such that a minimum cut will correspond to a maximum utility bond. This idea follows an approach used, for example, in the Project Selection Problem [10].

In order to construct $G_{extended}^{\mathcal{C}}$ we add a new source s and a new sink t to the graph $G^{\mathcal{C}}$. For each node $n \in N_C^+$ we add an edge (s, n) with capacity $u(n)$. For each node $m \in N_C^-$ we add an edge (m, t) with capacity $-u(m)$ (thus a positive utility). The rest of capacities (the capacities of edges corresponding to those in $G^{\mathcal{C}}$ are set to ∞.

Example 3. Consider the bipolar argument framework whose graph is shown in Figure 3. The corresponding $G_{extended}^{\mathcal{C}}$ for the coalition $\{a, b, c, d, e, f\}$ is shown in Figure 4. For readability, we have omitted the "∞" label on edges $\{(a, d), (a, e), (b, e), (b, f), (c, f)\}$.

Theorem 1. *If (A', B') is a minimum cut in $G_{extended}^{\mathcal{C}}$ then the set $A = A' - \{s\}$ is a maximum utility bond.*

Proof. The capacity of the cut $(\{s\}, \mathcal{C} \cup \{t\})$ is $C = \sum_{n \in N_C^+} u(n)$. So, the maximum flow value in this network is at most C.

We want to ensure that if (A', B') is a minimum cut in the graph $G_{extended}^{\mathcal{C}}$, then $A = A' - \{s\}$ satisfies the bond property (that is, it contains all of the

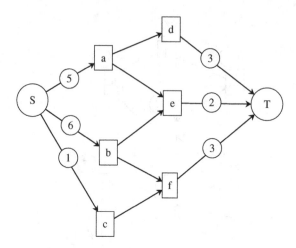

Fig. 4. Extended network of a coalition in a Bipolar Argumentation Framework

supported elements). Therefore, if the node $i \in A$ has an edge (i, j) in the graph then we must have $j \in A$. Since the capacities of all the edges coming from the graph $G^{\mathcal{C}}$ have capacity ∞ this means that we cannot cut along such edge (the flow would be ∞).

Therefore, if we compute a minimum cut (A', B') in $G^{\mathcal{C}}_{extended}$ we have that $A' - \{s\}$ is a bond. We now prove that it is of maximum utility.

Let us consider any bond $\mathcal{B}$. Let $A' = \mathcal{B} \cup \{s\}$ and $B' = (\mathcal{C} - \mathcal{B}) \cup \{t\}$ and consider the cut (A', B').

Since $\mathcal{B}$ is a bond, no edge (i, j) crosses this cut (if not there will be a supported argument not in $\mathcal{B}$). The capacity of the cut (A', B') satisfying the bond support constraints as defined from $\mathcal{C}$ is $c(A', B') = C - \sum_{a \in \mathcal{C}} u(a)$, where $C = \sum_{a \in \mathcal{C}} u(a)$. We can now prove that the minimum cut in G' determines the bond of maximum utility. The cuts (A', B') of capacity at most C are in a one-to-one correspondence with bonds $A = A' - \{s\}$. The capacity of such a cut is:

$$c(A', B') = C - u(A)$$

The capacity value is a constant, independent of the cut, so the cut with the minimum capacity corresponds to maximum utility bonds.

We have thus proved a polynomial time algorithm for the maximum utility bond decision. While this seems like a strong result, it should be noted that we made use of the fact that our input was a coalition rather than the full argument system. Typically, agents must consider the entire set of arguments and must therefore identify the coalitions themselves. Doing so is an NP-complete problem [3].

To conclude, we discuss the complexity of the two decision problems (bond finding and coalition finding) in an abstract setting, where the input is a bipolar argument framework (as opposed to a compiled knowledge base such as ASPIC+ permits, though this results in an exponential blow-up of the size of the domain).

P1

Input: A bipolar argumentation framework, an utility function and $k \in \mathbf{Z}$.
Question: Is there a coalition with the utility $\geq$ k.

P2

Input: A bipolar argumentation framework, an utility function and $k \in \mathbf{Z}$.
Question: Is there a bond with the utility $\geq$ k.

Both problems are NP-complete.

Proof. (*Sketch*) Clearly, the two problems belongs to NP. To prove the NP-completeness, let us consider an undirected graph G and a utility function defined on its nodes. If the edges of this graph are considered directed (thus obtaining a digraph that will correspond to the attack digraph), and each non-edge in G is replaced by a pair of opposed directed support edges, the coalitions in the Bipolar Argumentation Framework obtained are exactly the maximal stables in the initial graph. Indeed, these sets are conflict-free in the Bipolar Argumentation Framework and clearly connected in the support digraph. Thus, deciding if a given undirected graph with weights on its vertices has a maximal stable set of total weight greater or equal than some threshold can be reduced to **P1** (or **P2**). Since this problem is NP-complete [4] we have that **P1** and **P2** are NP-complete.

4 Discussion and Conclusions

In this paper we have introduced the notion of a maximum utility bond. These bonds are related to the justification of arguments when considered in the context of a bipolar argument framework. Since there can be many argument justifications, one needs to identify a heuristic for computing the "best" justification when arguing. We considered a utility based heuristic in which we have assumed that each argument has associated either a positive or a negative numerical utility. Such utility could correspond to the information revealing cost of the argument, or the degree of confidence the agent has in the argument, etc. Furthermore we have assumed that the utility function is additive. We have then described a polynomial time algorithm for computing maximum utility bonds, assuming that coalitions have already been identified.

In the future, we intend to investigate two significant extensions to the current work. First, we have assumed that utilities are additive. However, this simplification is not — in practice — realistic. The presence, or absence, of combinations of arguments could result in very different utilities, such as in the case where revealing some secret information makes other secret information unimportant to keep hidden. To address this case, we can transform our problem into a multi-agent resource allocation (MARA) problem, where arguments are transformed into resources. We must potentially consider an exponentially large input domain (i.e. all possible combinations of arguments), and in [8] such an exponential input was dealt with in the context of coalition formation. We therefore intend to apply their techniques, noting that additional constraints must be introduced on the technique's outputs to capture the nature of our domain.

Another potential avenue of future work arises by noting that we have implicitly combined the cost of revealing an argument, and attacks due to the argument on a bond into a single number (the negative utility). In reality, these two values are different; by separating them out, we could perform a minimisation that reflects the different potential preferences of a reasoner.

References

1. Amgoud, L., Cayrol, C., Lagasquie-Schiex, M.C., Livet, P.: On bipolarity in argumentation frameworks. International Journal of Intelligent Systems 23(10), 1062–1093 (2008)
2. Amgoud, L., Prade, H.: Reaching agreement through argumentation: a possiblistic approach. In: Proceedings of the 9th International Conference on the Principles of Knowledge Representation and Reasoning, pp. 175–182 (2004)
3. Cayrol, C., Lagasquie-Schiex, M.-C.: Coalitions of arguments: A tool for handling bipolar argumentation frameworks. International Journal of Intelligent Systems 25(1), 83–109 (2010)
4. Garey, M.R., Johnson, D.S.: Computers and Intractability: A Guide to the Theory of NP-completeness. Freeman and Co. (1979)
5. Goldberg, V., Tarjan, R.: A new approach to the maximum flow problem. In: Proceedings of the Eighteenth Annual ACM Symposium on Theory of Computing, pp. 136–146 (1986)
6. Kakas, A.C., Maudet, N., Moraïtis, P.: Layered strategies and protocols for argumentation-based agent interaction. In: Rahwan, I., Moraïtis, P., Reed, C. (eds.) ArgMAS 2004. LNCS (LNAI), vol. 3366, pp. 64–77. Springer, Heidelberg (2005)
7. Oren, N., Norman, T.J., Preece, A.D.: Arguing with confidential information. In: Proc. of ECAI 2006, pp. 280–284 (2006)
8. Rahwan, T., Michalak, T.P., Croitoru, M., Sroka, J., Jennings, N.R.: A network flow approach to coalitional games. In: ECAI. Frontiers in Artificial Intelligence and Applications, vol. 215, pp. 1017–1018. IOS Press (2010)
9. Rienstra, T., Thimm, M., Oren, N.: Opponent models with uncertainty for strategic argumentation. In: Proceedings of the 23rd International Joint Conference on Artificial Intelligence (to appear, 2013)
10. Tardos, E., Kleinberg, J.: Algorithm Design. Addison-Wesley (2005)

Different Classes of Graphs to Represent Microstructures for CSPs*

Achref El Mouelhi, Philippe Jégou, and Cyril Terrioux

LSIS - UMR CNRS 7296
Aix-Marseille Université
Avenue Escadrille Normandie-Niemen
13397 Marseille Cedex 20, France
{achref.elmouelhi,philippe.jegou,cyril.terrioux}@lsis.org

Abstract. The CSP formalism has shown, for many years, its interest for the representation of numerous kinds of problems, and also often provide effective resolution methods in practice. This formalism has also provided a useful framework for the knowledge representation as well as to implement efficient methods for reasoning about knowledge. The data of a CSP are usually expressed in terms of a constraint network. This network is a (constraints) graph when the arity of the constraints is equal to two (binary constraints), or a (constraint) hypergraph in the case of constraints of arbitrary arity, which is generally the case for problems of real life. The study of the structural properties of these networks has made it possible to highlight certain properties, which led to the definition of new tractable classes, but in most cases, they have been defined for the restricted case of binary constraints. So, several representations by graphs have been proposed for the study of constraint hypergraphs to extend the known results to the binary case. Another approach, finer, is interested in the study of the microstructure of CSP, which is defined by graphs. This helped, offering a new theoretical framework to propose other tractable classes.

In this paper, we propose to extend the notion of microstructure to any type of CSP. For this, we propose three kinds of graphs that can take into account the constraints of arbitrary arity. We show how these new theoretical tools can already provide a framework for developing new tractable classes for CSPs. We think that these new representations should be of interest for the community, firstly for the generalization of existing results, but also to obtain original results.

1 Preliminaries

Constraint Satisfaction Problems (CSPs, see [1] for a state of the art) provide an efficient way of formulating problems in computer science, especially in Artificial Intelligence. Formally, a *constraint satisfaction problem* is a triple (X, D, C),

---

* This work was supported by the French National Research Agency under grant TUPLES (ANR-2010-BLAN-0210).

M. Croitoru et al. (Eds.): GKR 2013, LNAI 8323, pp. 21–38, 2014.

where $X = \{x_1, \ldots, x_n\}$ is a set of n variables, $D = (D_{x_1}, \ldots, D_{x_n})$ is a list of finite domains of values, one per variable, and $C = \{C_1, \ldots, C_e\}$ is a finite set of e constraints. Each constraint C_i is a pair $(S(C_i), R(C_i))$, where $S(C_i) = \{x_{i_1}, \ldots, x_{i_k}\} \subseteq X$ is the *scope* of C_i, and $R(C_i) \subseteq D_{x_{i_1}} \times \cdots \times D_{x_{i_k}}$ is its *compatibility relation*. The *arity* of C_i is $|S(C_i)|$.

We assume that each variable appears at least in the scope of one constraint and that the relations are represented in extension (e.g. by providing the list of allowed tuples) even if for some parts of this work, this hypothesis is not required. A CSP is called *binary* if all constraints are of arity 2 (we denote C_{ij} the binary constraint whose scope is $S(C_{ij}) = \{x_i, x_j\}$). Otherwise, if the constraints are of arbitrary arity, a CSP is said to be *non binary*. The structure of the constraint network is represented by the hypergraph (X, C) (which is a graph in the binary case) whose vertices correspond to variables and edges to the constraint scopes. An assignment on a subset of X is said to be *consistent* if it does not violate any constraint.

Testing whether a CSP has a *solution* (i.e. a consistent assignment on all the variables) is known to be NP-complete. So, many works have been realized to make the solving of instances more efficient by using optimized backtracking algorithms, filtering techniques based on constraint propagation, heuristics...

Another way is related to the study of tractable classes defined by properties of constraint networks. E.g., it has been shown that if the structure of this network, that is a graph for binary CSPs, is acyclic, it can be solved in linear time [2]. This kind of result has been extended to hypergraphs in [3,4]. Using these theoretical results, some practical methods to solve CSPs have been defined, such as Tree-Clustering [5] which can be efficient in practice [6]. So, the study of such properties for graphs or hypergraphs has shown its interest regarding the constraint network.

Graphs properties have also been exploited to study the properties of compatibility relations for the case of binary CSPs. This is made possible thanks to a representation called microstructure that we can associate to a binary CSP. A microstructure is defined as follows:

Definition 1 (Microstructure). *Given a binary CSP $P = (X, D, C)$, the microstructure of P is the undirected graph $\mu(P) = (V, E)$ with:*

- $V = \{(x_i, v_i) : x_i \in X, v_i \in D_{x_i}\}$,
- $E = \{ \{(x_i, v_i), (x_j, v_j)\} \mid i \neq j, C_{ij} \notin C \text{ or } C_{ij} \in C, (v_i, v_j) \in R(C_{ij})\}$

The transformation of a CSP instance using this representation can be considered as a reduction from the CSP problem to the well known CLIQUE problem [7] seeing that it can be realized in polynomial time and using the theorem [8] recalled below:

Theorem 1. *An assignment of variables in a binary CSP P is a solution iff this assignment is a clique of size n (the number of variables) in $\mu(P)$.*

The interest to consider the microstructure was firstly shown in [8] in order to detect new tractable classes for CSP based on Graph Theory. Indeed, while determining whether the microstructure contains a clique of size n is NP-complete, this task can be achieved, in some cases, in polynomial time. For example, using a famous result of Gavril [9], Jégou has shown that if the microstructure of a binary CSP is triangulated, then this CSP can be solved in polynomial time. By this way, a new tractable class for binary CSPs has been defined since it is also possible to recognize triangulated graphs in polynomial time.

Later, in [10], applying the same approach and also [9], Cohen shows that the class of binary CSPs with triangulated complement of microstructure is tractable, the achievement of arc-consistency being a decision procedure.

More recently, other works have defined new tractable classes of CSPs thanks to the study of microstructure. For example, generalizing the result on triangulated graphs, [11] have shown that the class of binary CSPs the microstructure of which is a *perfect graph* constitutes also a tractable class. Then, in [12], the class BTP, which is defined by forbidden patterns (as for triangulated graphs), has been introduced. After that, [13] also exploit the microstructure, but in another way, by presenting new results on the effectiveness of classical algorithms for solving CSPs when the number of maximal cliques in the microstructure of binary CSPs is bounded by a polynomial.

The study of the microstructure has also shown its interest in other fields. For example, for the problem of counting the number of solutions [14], or for the study of symmetries in binary CSPs [15,16]. Thus, the microstructure appears as an interesting tool for the study of CSPs, or more precisely, for the theoretical study of CSPs.

This notion has been studied and exploited in the limited field of binary CSPs, even if the microstructure for non binary CSPs has already been considered. Indeed, in [10], the complement of the microstructure of a non binary CSP is defined as a hypergraph:

Definition 2 (Complement of the Microstructure). *Given a binary CSP* $P = (X, D, C)$, *the* Complement of the Microstructure *of P is the hypergraph* $\overline{\mathcal{M}}(P) = (V, E)$ *such that:*

- $V = \{(x_i, v_i) : x_i \in X, v_i \in D_{x_i}\}$,
- $E = E_1 \cup E_2$ *such that*
    - $E_1 = \{ \{(x_i, v_j), (x_i, v_{j'})\} \mid x_i \in X \text{ and } j \neq j'\}$
    - $E_2 = \{\{(x_{i_1}, v_{i_1}), \ldots (x_{i_k}, v_{i_k})\} \mid C_i \in C, S(C_i) = \{x_{i_1}, \ldots, x_{i_k}\} \text{ and } (v_{i_1}, \ldots v_{i_k}) \notin R(C_i)\}$

One can see that for the case of binary CSPs, this definition is a generalization of the microstructure since the Complement of the Microstructure is then exactly the *complement of the graph* of microstructure. Unfortunately, while it is easily possible to consider the complement of a graph, for hypergraphs this notion is not clearly defined in Hypergraph Theory. For example, should we consider all possible hyperedges of the hypergraph (i.e. all the subsets of V) by associating to each one a universal relation? In this case, the size of representation would be potentially exponential w.r.t. the size of the considered instance of CSP. As a consequence, the notion of microstructure for non binary CSPs is not explicitly defined in [10], and to our knowledge, this question seems to be considered as open today. Moreover, to our knowledge, it turns out that this definition of complement of the microstructure has not really been exploited for non binary CSPs, even in the paper where it is defined since [10] only exploits it for binary CSPs. More generally, exploiting a definition of a microstructure based on hypergraphs seems to be really more difficult than when it is defined by graphs. Indeed, it is well known that the literature of Graph Theory is really more extended than one of Hypergraph Theory. So, the theoretical results and efficient algorithms to manage them are more numerous, offering a larger number of existing tools which can be operated for graphs rather than for hypergraphs.

So, in this paper, to extend this notion to CSPs with constraints of arbitrary arity, we propose another way than the one introduced in [10]. We propose to preserve the graph representation rather than the hypergraph representation. This is possible using known representations of constraint networks by graphs. So, we introduce three possible microstructures, based on the *dual representation*, on the *hidden variable representation* and on the *mixed encoding* [17] of non binary CSPs. We study the basic properties of such microstructures. We also give some possible tracks to exploit these microstructures for future theoretical developments, focusing particularly on extensions of tractable classes to non binary CSPs.

The next section introduces different possibilities of microstructures for non binary CSPs while the third section shows some first results exploiting them. The last section presents a conclusion.

2   Microstructures for Non Binary CSPs

As indicated above, the first evocation of the notion of microstructure to non-binary CSPs was proposed by Cohen in [10] and is based on hypergraphs. In contrast, we will propose several microstructures based on graphs. To do this, we will rely on the conversion of non-binary CSPs to binary CSPs. The well known methods are the dual encoding (also called dual representation), the hidden transformation (also called hidden variable representation) and the mixed encoding (also called combined encoding).

2.1   Microstructure Based on Dual Representation

The dual encoding appeared in CSPs in [18]. It is based on the graph representation of hypergraphs called *Line Graphs* which has been introduced in the (Hyper)Graph Theory and which are called *Dual Graphs* for CSPs. This representation was also used before in the field of Relational Database Theory (Dual Graphs were called *Qual Graphs* in [19]). In this encoding, the constraints of the original problem become the variables (also called *dual variables*). The domain of each new variable is exactly the set of tuples allowed by the original constraint. Then a binary constraint links two dual variables if the original constraints share at least one variable (i.e. the intersection between their scopes is not empty). So, this representation allows to define a binary instance of CSP which is equivalent to the considered non binary instance.

Definition 3 (Dual Representation). *Given a CSP* $P = (X, D, C)$, *the Dual Graph* (C, F) *of* (X, C) *is such that* $F = \{\{C_i, C_j\} : S(C_i) \cap S(C_j) \neq \emptyset\}$. *The Dual Representation of* P *is the CSP* (C_D, R_D, F_D) *such that:*

- $C_D = \{S(C_i) : C_i \in C\}$,
- $R_D = \{R(C_i) : C_i \in C\}$
- $F_D = \{F_k : S(F_k) \in F$ *and for* $S(F_k) = \{C_i, C_j\}, R(F_k) = \{(t_i, t_j) \in R(C_i) \times R(C_j) : t_i[S(C_i) \cap S(C_j)] = t_j[S(C_i) \cap S(C_j)]\}\}$.

The associated microstructure is then immediately obtained considering the microstructure of this equivalent binary CSP:

Definition 4 (DR-Microstructure). *Given a CSP* $P = (X, D, C)$ *(not necessarily binary), the* Microstructure based on Dual Representation *of* P *is the undirected graph* $\mu_{DR}(P) = (V, E)$ *such that:*

- $V = \{(C_i, t_i) : C_i \in C, t_i \in R(C_i)\}$,
- $E = \{\ \{(C_i, t_i), (C_j, t_j)\}\ |\ i \neq j, t_i[S(C_i) \cap S(C_j)] = t_j[S(C_i) \cap S(C_j)]\}$

where $t[Y]$ *denotes the restriction of* t *to the variables of* Y.

Note that this definition has firstly been introduced in [20]. As for the microstructure, there is a direct relationship between cliques and solutions of CSPs:

Theorem 2. *A CSP* P *has a solution iff* $\mu_{DR}(P)$ *has a clique of size* e *(the number of constraints).*

Proof: By construction, $\mu_{DR}(P)$ is e-partite, and any clique contains at most one vertex (C_i, t_i) per constraint $C_i \in C$. Hence each e-clique of $\mu_{DR}(P)$ has exactly one vertex (C_i, t_i) per constraint $C_i \in C$. By construction of $\mu_{DR}(P)$, any two vertices $(C_i, t_i), (C_j, t_j)$ joined by an edge (in particular, in some clique) satisfy $t_i[S(C_i) \cap S(C_j)] = t_j[S(C_i) \cap S(C_j)]$. Hence all the tuples t_i in a clique join together, and it follows that the e-cliques of $\mu_{DR}(P)$ correspond exactly to tuples t which are joins of one allowed tuple per constraint, that is, to solutions of P.   □

Consider the example 1 which will be used in this paper:

Example 1. $P = (X, D, C)$ *has five variables* $X = \{x_1, \ldots, x_5\}$ *with domains* $D = \{D_{x_1} = \{a, a'\}, D_{x_2} = \{b\}, D_{x_3} = \{c\}, D_{x_4} = \{d, d'\}, D_{x_5} = \{e\}\}$. $C = \{C_1, C_2, C_3, C_4\}$ *is a set of four constraints with* $S(C_1) = \{x_1, x_2\}$, $S(C_2) = \{x_2, x_3, x_5\}$, $S(C_3) = \{x_3, x_4, x_5\}$ *and* $S(C_4) = \{x_2, x_5\}$. *The relations associated to the previous constraints are given by these tables:*

$R(C_1)$	
x_1	x_2
a	b
a'	b

$R(C_2)$		
x_2	x_3	x_5
b	c	e

$R(C_3)$		
x_3	x_4	x_5
c	d	e
c	d'	e

$R(C_4)$	
x_2	x_5
b	e

The DR-Microstructure of this example is shown in figure 1. We have 4 constraints, then $e = 4$. Thanks to Theorem 2, a solution of P is a clique of size 4, e.g. $\{ab, bce, be, cde\}$ (in the examples, we denote directly t_i the vertex (C_i, t_i) and v_j the vertex (x_j, v_j) when there is no ambiguity).

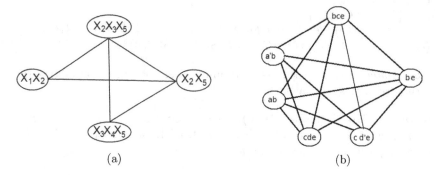

(a)   (b)

Fig. 1. Dual Graph (a) and DR-Microstructure (b) of the CSP of the example 1

Assuming that relations of instances are given by tables (it will be the same for next microstructures), the size of the DR-Microstructure is bounded by a polynomial in the size of the CSP, since $|E| \leq |V|^2$ with $|V| = \Sigma_{C_i \in C} |\{t_i \in R(C_i)\}|$. Moreover, given an instance of CSP, computing its DR-Microstructure can be achieved in polynomial time.

More generally, with a similar approach, one could define a set of DR-Microstructures for a given non binary CSP. Indeed, it is known that for some CSPs, some edges of their dual representation can be deleted, while preserving the equivalence (this question has been studied in [21]). Before, in [4], it has been shown that given a hypergraph, we can define a collection of dual (or qual) subgraphs deleting edges while preserving the connexity between shared variables. Some of these subgraphs being minimal for inclusion and also for the number of edges. These graphs can be called *Qual Subgraphs* while the minimal ones are called *Minimal Qual Graphs*.

Applying this result, in [21], it is shown that for a given non binary CSP, there is a collection of equivalent binary CSPs (the maximal one being its dual encoding), assuming that their associated graphs preserve the connexity.

Definition 5 (Dual Subgraph Representation). *Given a CSP $P = (X, D, C)$ and a Dual Graph (C, F) of (X, C), a Dual Subgraph (C, F') of (C, F) is such that $F' \subseteq F$ and $\forall C_i, C_j \in C$ such that $S(C_i) \cap S(C_j) \neq \emptyset$, there is a path $(C_i = C_{k_1}, C_{k_2}, \dots C_{k_l} = C_j)$ in (C, F') such that $\forall u, 1 \leq u < l, S(C_i) \cap S(C_j) \subseteq S(C_{k_u}) \cap S(C_{k_{u+1}})$.*

A *Dual Subgraph Representation of* P *is the CSP* (C_D, R_D, F'_D) *such that* $F'_D \subseteq F_D$ *where* (C_D, R_D, F_D) *is the Dual Representation of* P.

Figure 2 represents the two Dual Subgraphs of the Dual Graph given in the figure 1. We can see that despite the deletion of one edge in each subgraph, the connection between vertices containing x_2 is preserved by the existence of appropriate paths. In the first case, the connection between C_1 and C_2 is preserved by the path (C_1, C_4, C_2) while in the second case, the connection between C_1 and C_4 is preserved by the path (C_1, C_2, C_4).

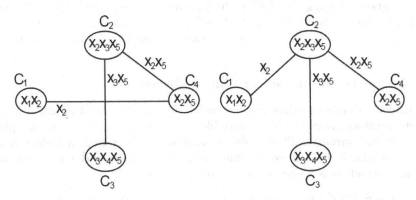

Fig. 2. The two Dual Subgraphs of the Dual Graph of the figure 1 (a)

The equivalence between the Dual Subgraph Representation and the non binary CSP is given by the next theorem [21]:

Theorem 3. *There is a bijection between the set of solutions of a CSP* $P = (X, D, C)$ *and set of solutions of its Dual Subgraph Representation* (C_D, R_D, F'_D).

So, considering these subgraphs, we can extend the previous definition of DR-Microstructures:

Definition 6 (DSR-Microstructure). *Given a CSP* $P = (X, D, C)$ *(not necessarily binary) and one of its Dual Subgraph* (C, F'), *the Microstructure based on Dual Subgraph Representation of* P *is the undirected graph* $\mu_{DSR}(P, (C, F'))$ $= (V, E)$ *with:*

- $V = \{(C_i, t_i) : C_i \in C, t_i \in R(C_i)\}$,
- $E = E_1 \cup E_2$ *such that*
  - $E_1 = \{ \{(C_i, t_i), (C_j, t_j)\} \mid \{(C_i, C_j)\} \in F', t_i[S(C_i) \cap S(C_j)] = t_j[S(C_i) \cap S(C_j)]\}$
  - $E_2 = \{ \{(C_i, t_i), (C_j, t_j)\} \mid \{(C_i, C_j)\} \notin F'\}$.

With this representation, we have the same kind of properties since the size of the DSR-Microstructure is bounded by the same polynomial in the size of the CSP as for DR-Microstructure. Moreover, the computing of the DSR-Microstructure can be achieved in polynomial time. Nevertheless, while Dual Subgraphs are subgraphs of Dual Graph, the DR-Microstructure is a subgraph of the DSR-Microstructure since for each deleted edge, a universal binary relation needs to be considered. Note that the property about the cliques is preserved:

Theorem 4. *A CSP P has a solution iff $\mu_{DSR}(P, (C, F'))$ has a clique of size e.*

Proof: Using the theorem 3, we know that all Dual Subgraph Representations of a CSP P have the same number of solutions as P. Moreover, since $\mu_{DR}(P)$ is a partial graph of $\mu_{DSR}(P, (C, F'))$ which is an e-partite graph, each e-clique of $\mu_{DR}(P)$ is also a e-clique of $\mu_{DSR}(P, (C, F'))$, and thus, there is no more e-clique in $\mu_{DSR}(P, (C, F'))$. So, it is sufficient to use theorem 2 to obtain the result. □

2.2   Microstructure Based on Hidden Variable

The hidden variable encoding was inspired by Peirce [22] (cited in [23]). In the hidden transformation, the set of variables contains the original variables plus the set of dual variables. Then a binary constraint links a dual variable and an original variable if the original variable belongs to the scope of the dual variable. The microstructure is based on this binary representation:

Definition 7 (HT-Microstructure). *Given a CSP $P = (X, D, C)$ (not necessarily binary), the Microstructure based on Hidden Transformation of P is the undirected graph $\mu_{HT}(P) = (V, E)$ with:*

- *$V = S_1 \cup S_2$ such that*
  - *$S_1 = \{(x_i, v_i) : x_i \in X, v_i \in D_{x_i}\}$,*
  - *$S_2 = \{(C_i, t_i) : C_i \in C, t_i \in R(C_i)\}$,*
- *$E = \{ \{(C_i, t_i), (x_j, v_j)\} \mid either\ x_j \in S(C_i)\ and\ v_j = t_i[x_j]\ or\ x_j \notin S(C_i)\}$.*

Figures 3 and 4 represent respectively the hidden graph and the HT-Microstructure based on the hidden transformation for the CSP of example 1. We can see that the HT-Microstructure is a bipartite graph. This will affect the

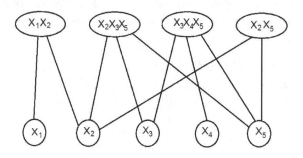

Fig. 3. Hidden graph of the CSP of the example 1

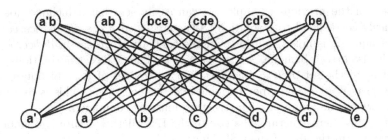

Fig. 4. HT-Microstructure of the CSP of the example 1

representation of solutions. Before that, we should recall that a *biclique* is a complete bipartite subgraph, i.e. a bipartite graph in which every vertex of the first set is connected to all vertices of the second set. A biclique between two subsets of vertices of sizes i and j is denoted $K_{i,j}$. The solutions will correspond to some particular bicliques:

Lemma 1. *In a HT-Microstructure, a $K_{n,e}$ biclique with e tuples, such that no two tuples belong to the same constraint, cannot contain two different values of the same variable.*

Proof: We assume that a $K_{n,e}$ biclique with e tuples, such that no two tuples belong to the same constraint, can contain two different values v_j and v'_j of the same variable x_j. Therefore, there is at least one constraint C_i such that $x_j \in S(C_i)$. Thus, $t_i[x_j] = v_j$, v'_j or another v''_j. Hence, in all three cases, we have a contradiction since t_i cannot be connected to two different values of the same variable.　　　□

Lemma 2. *In a HT-Microstructure, a $K_{n,e}$ biclique with n values, such that no two values belong to the same variable, cannot contain two different tuples of the same constraint.*

Proof: We assume that a $K_{n,e}$ biclique with n values, such that no two values belong to the same variable, can contain two different tuples t_i and t'_i of the same constraint C_i. Therefore, there is at least one variable x_j such that $t_i[x_j] \neq t'_i[x_j]$. If $v_j = t_i[x_j]$ and $v'_j = t'_i[x_j]$ belong both to the $K_{n,e}$ biclique, we have a contradiction since we cannot have two values of the same variable.　　□

Using these two lemmas, since a $K_{n,e}$ biclique with n values and e tuples such that no two values belong to the same variable and no two tuples belong to the same constraint corresponds to an assignment on all the variables which satisfies all the constraints, we can deduce the following theorem:

Theorem 5. *Given a CSP $P = (X, D, C)$ and $\mu_{HT}(P)$ its HT-Microstructure, P has a solution iff $\mu_{HT}(P)$ has a $K_{n,e}$ biclique with n values and e tuples such that no two values belong to the same domain and no two tuples belong to the same constraint.*

Based on the previous example, we can easily see that a biclique does not necessarily correspond to a solution. Although $\{a, a', b, c, e, ab, ab', bce, be\}$ is a $K_{5,4}$ biclique, it is not a solution. On the contrary, $\{a, b, c, d, e, ab, bce, be, cde\}$ which is also a $K_{5,4}$ biclique, is a solution of P. Then, the set of solutions is not equivalent to the set of $K_{n,e}$ bicliques, but to the set of $K_{n,e}$ bicliques which contain exactly one vertex per variable and per constraint. This is due to the manner which the graph of microstructure must be completed.

As for DR-Microstructure, the size of the HT-Microstructure is bounded by a polynomial in the size of the CSP, since:

- $|V| = \Sigma_{x_i \in X}|D_{x_i}| + \Sigma_{C_i \in C}|\{t_i \in R(C_i)\}|$ and
- $|E| \leq \Sigma_{x_i \in X}|D_{x_i}| \times \Sigma_{C_i \in C}|\{t_i \in R(C_i)\}|$.

Moreover, given an instance of CSP, computing its HT-Microstructure can also be achieved in polynomial time.

For the third microstructure, we propose another manner to complete the graph of microstructure: this new way of representation is also deduced from hidden encoding.

2.3   Microstructure Based on Mixed Encoding

The Mixed Encoding [17] of non binary CSPs uses at the same time dual encoding and hidden variable encoding. This approach allows us to connect the values of dual variables to the values of original variables, two tuples of two different constraints and two values of two different variables. More precisely:

Definition 8 (ME-Microstructure). *Given a CSP $P = (X, D, C)$ (not necessarily binary), the* Microstructure based on Mixed Encoding *of P is the undirected graph $\mu_{ME}(P) = (V, E)$ with:*

- $V = S_1 \cup S_2$ *such that*
    - $S_1 = \{(C_i, t_i) : C_i \in C, t_i \in R(C_i)\}$,
    - $S_2 = \{(x_j, v_j) : x_j \in X, v_j \in D_{x_j}\}$,
- $E = E_1 \cup E_2 \cup E_3$ *such that*
    - $E_1 = \{ \{(C_i, t_i), (C_j, t_j)\} \mid i \neq j, t_i[S(C_i) \cap S(C_j)] = t_j[S(C_i) \cap S(C_j)]\}$
    - $E_2 = \{ \{(C_i, t_i), (x_j, v_j)\} \mid either \ x_j \in S(C_i) \ and \ v_j = t_i[x_j] \ or \ x_j \notin S(C_i)\}$
    - $E_3 = \{ \{(x_i, v_i), (x_j, v_j)\} \mid x_i \neq x_j\}$.

The microstructure based on the mixed encoding of the CSP of example 1 is shown in figure 6 while figure 5 represents the corresponding mixed graph. We can observe that in this encoding, we have the same set of vertices as for the HT-Microstructure while for edges, we have the edges which belong to the DR-Microstructure and the HT-Microstructure, plus all the edges between values of domains that could appear in the classical microstructure of binary CSPs. This will have an impact on the relationship between the solutions of the CSP and the properties of the graph of ME-Microstructure. The next lemma formalizes these observations:

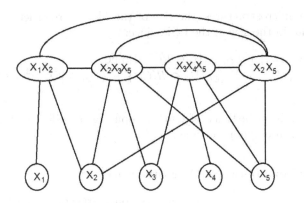

Fig. 5. Mixed graph of the CSP of the example 1

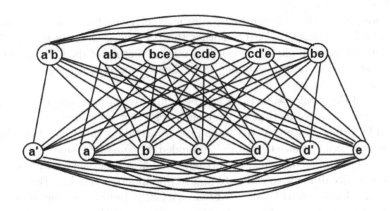

Fig. 6. ME-Microstructure (b) of the CSP of the example 1

Lemma 3. *In a ME-Microstructure, a clique on $n + e$ vertices cannot contain two different values of the same variable, neither two different tuples of the same constraint.*

Proof: Let v_i and v_i' be two values of the same variable x_i. By definition, the vertices corresponding to v_i and v_i' cannot be adjacent and so cannot belong to the same clique. Likewise, for the tuples. □

According to this lemma, there is a strong relationship between cliques and solutions of CSPs:

Theorem 6. *A CSP P has a solution iff $\mu_{ME}(P)$ has a clique of size $n + e$.*

Proof: In a ME-Microstructure, according to lemma 3, a clique on $n+e$ vertices contains exactly one vertex per variable and per constraint. So it corresponds to an assignment of n variables which satisfies e constraints, i.e. a solution of P. □

As for other microstructures, the size of the ME-Microstructure is bounded by a polynomial in the size of the CSP, since:

- $|V| = \Sigma_{x_i \in X}|D_{x_i}| + \Sigma_{C_i \in C}|\{t_i \in R(C_i)\}|$ and
- $|E| \leq \Sigma_{x_i \in X}|D_{x_i}| \times \Sigma_{C_i \in C}|\{t_i \in R(C_i)\}| + (\Sigma_{x_i \in X}|D_{x_i}|)^2 + (\Sigma_{C_i \in C}|\{t_i \in R(C_i)\}|)^2$.

Moreover, given an instance of CSP, computing its ME-Microstructure can also be achieved in polynomial time.

2.4 Comparisons between Microstructures

Firstly, we must observe that none of these microstructures can be considered as a generalization of the classical microstructure of binary CSPs. Indeed, given a binary CSP P, we have $\mu(P) \neq \mu_{DR}(P)$ (and $\mu(P) \neq \mu_{DSR}(P)$), $\mu(P) \neq \mu_{HT}(P)$ and $\mu(P) \neq \mu_{ME}(P)$.

Moreover, while the DR-Microstructure is exactly the binary microstructure of the dual CSP (idem for DSR), neither the HT-Microstructure nor the ME-Microstructure correspond to the classical microstructure of the CSP associated to the binary representations coming from the original instance, because of the way to complete these graphs.

Finally, all these microstructures can be computed in polynomial time. This is true because we assume that compatibility relations associated to constraints are given by tables. Note that the same property holds without this hypothesis, but assuming that the size of scopes is bounded by constants, since we consider here CSPs with finite domains. Nevertheless, from a practical viewpoint, they seem to be really difficult to compute and to manipulate explicitly. But it is the same for the classical microstructure of binary CSPs. Indeed, this should require having relations given by tables or to compute all the satisfying tuples. And even if this is the case, except for small instances, this would lead generally to build graphs with a too large number of edges. However, this last point is not really a problem because our motivation in this paper concerns the proposal of new tools for the theoretical study of non binary CSPs. To this end, the following section presents some first results exploiting these microstructures for defining new tractable classes.

3   Some Results Deduced from Microstructures

We now present some results which can be deduced from the analysis of these microstructures. For this, we will study three tractable classes, including those corresponding to well known properties as "0-1-all" [24] and BTP [12] for which it is necessary to make a distinctness between the vertices in the graph, and a third one for which the vertices do not have to be distinguished.

3.1 Microstructures and Number of Cliques

In [13], it is shown that if the number of maximal cliques in the microstructure of a binary CSP (denoted $\omega_\#(\mu(P))$) is bounded by a polynomial, then classical algorithms like Backtracking (BT), Forward Checking (FC [25]) or Real Full Look-ahead (RFL [26]) solve the corresponding CSP in polynomial time. Exactly, the cost is bounded by $O(n^2 d \cdot \omega_\#(\mu(P)))$ for BT and FC, and by $O(ned^2 \cdot \omega_\#(\mu(P)))$ for RFL. We analyze here if this kind of result can be extended to non binary CSPs, exploiting the different microstructures.

More recently in [20], these results have been generalized to non binary CSPs, exploiting the Dual Representation, using the algorithms nBT, nFC and nRFL, which are the non binary versions of BT, FC and RFL. More precisely, by exploiting a particular ordering for the assignment of variables, it is shown that the complexity is bounded by $O(nea \cdot d^a \cdot \omega_\#(\mu_{DR}(P)))$ for nBT, and by $O(nea \cdot r^2 \cdot \omega_\#(\mu_{DR}(P)))$ for nFC and nRFL, where a is the maximum arity for constraints and r is the maximum number of tuples per compatibility relations.

Based on the time complexity of these algorithms, and regarding some classes of graphs with number of maximal cliques bounded by a polynomial, it is easy to define new tractable classes. Such classes of graph are, for example, *planar* graphs, *toroidal* graphs, graphs *embeddable in a surface* [27] or *CSG* graphs [28]. This result can be summarized by:

Theorem 7. *CSPs of arbitrary arities the DR-Microstructure of which is either a planar graph, a toroidal graph, a graph embeddable in a surface or a CSG graph, are tractable.*

For HT-Microstructures, such a result does not hold. Indeed, these microstructures are bipartite graphs. So the maximal cliques have size at most two since they correspond to edges and their number is the number of edges in the graph, which is then bounded by a polynomial, independently of the tractability of the instance.

For ME-Microstructures, such a result does not hold too, but for a different reason. By construction, the edges corresponding to the set $E_3 = \{ \{(x_i, v_i), (x_j, v_j)\} \mid x_i \neq x_j \}$ of definition 8 allow all the possible assignments of variables, making the number of maximal cliques exponential except for CSPs with a single value per domain.

3.2 Microstructures and BTP

The property BTP (Broken Triangle Property) [12] defines a new tractable class for binary CSPs while exploiting characteristics of the microstructure. The BTP class turns out to be important because it captures some tractable classes (such as the class of tree-CSPs and other semantical tractable classes such as RRM). The question is then: could we extend this property to non binary CSPs while exploiting characteristics of their microstructures? A first discussion about this appears in [12]. Here, we extend these works, by analyzing the question on the DR, HT and ME-Microstructures. Before, we recall the BTP property:

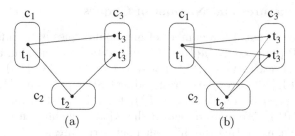

Fig. 7. DR-Microstructure of a non binary CSP satisfying BTP on its dual representation

Definition 9. *A CSP instance (X, D, C) satisfies the Broken Triangle Property (BTP) w.r.t. the variable ordering $<$ if, for all triples of variables (x_i, x_j, x_k) s.t. $x_i < x_j < x_k$, s.t. $(v_i, v_j) \in R(C_{ij})$, $(v_i, v_k) \in R(C_{ik})$ and $(v_j, v'_k) \in R(C_{jk})$, then either $(v_i, v'_k) \in R(C_{ik})$ or $(v_j, v_k) \in R(C_{jk})$. If none of these two tuples exist, (v_i, v_j), (v_i, v_k) and (v_j, v'_k) is called a Broken Triangle on x_k.*

In [12], it is shown that, if a binary CSP is BTP, finding a good ordering and solving it is feasible in $O(n^3 d^4 + ed^2) = O(n^3 d^4)$.

DR-Microstructure. To extend BTP to non binary CSPs, the authors propose to consider the Dual Graph as a binary CSP, translating directly the BTP property. We denote DBTP this extension. For example, Figure 7 presents the DR-Microstructure of an instance P involving three constraints. In Figure 7 (a), we can observe the presence of a broken triangle on c_3 if we consider the ordering $c_1 \prec c_2 \prec c_3$ and so P does not satisfy DBTP w.r.t. $\prec$. In contrast, in Figure 7(b), if either t_1 and t'_3 (blue edge) or t_2 and t_3 (red edge) are compatible, then P satisfies DBTP according to $\prec$.

But it is possible, analyzing the DR-Microstructure, to extend significantly the first results achieved in [12], these ones being limited to show that the binary tree-structured instances are BTP on their dual representation. For example, it can be shown that for binary CSPs, the properties of the classical microstructure are clearly different than the ones of the associated DR-Microstructure, proving that for a binary instance, the existence of broken triangles is not equivalent, considering one or the other of these two microstructures. Moreover, it can also be proved that if a non binary CSP has β-acyclic hypergraph [29,30], then its DR-Microstructure admits an order such that there is no broken triangle, thus satisfying BTP. Other results due to the properties of the DR-Microstructure can be deduced considering BTP. More details about these results are given in [31].

HT and ME-Microstructures. For HT-Microstructure, one can easily see that no broken triangle exist explicitly since this graph is bipartite. To analyze BTP on this microstructure, one should need to consider universal constraints (i.e. with universal relations) between vertices of the constraint graph resulting from Hidden Transformation. Also, we will directly study ME-microstructure because this microstructure has the same vertices as the HT-Microstructure and it has been completed with edges between these vertices. So, consider now the HT-Microstructure. Extending BTP on this microstructure is clearly more complicated because we must consider at least four different cases of triangles, because contrary to BTP or BTP on the DR-Microstructure, we have two kinds of vertices: tuples of relations and values of domains. Moreover, since for BTP, we must also consider orderings such as $i < j < k$, actually we must consider six kinds of triangles since it is possible, for BTP to permute the order of the two first variables: (1) $x_i < x_j < x_k$, (2) $x_i < x_j < C_k$, (3) $x_i < C_j < x_k$ (or $C_i < x_j < x_k$), (4) $x_i < C_j < C_k$ (or $C_i < x_j < C_k$), (5) $C_i < C_j < x_k$, (6) $C_i < C_j < C_k$.

One can notice the existence of a link for BTP between DR-Microstructure and ME-Microstructure. Indeed, if a non binary CSP P has a broken triangle on DR-Microstructure, for any possible ordering of the constraints, then P possesses necessarily a broken triangle for any ordering on mixed variables (variables and constraints). This leads us to the following theorem which seems to show the DR-Microstructure as the most promising one w.r.t. the BTP property:

Theorem 8. *If a CSP P satisfies BTP considering its ME-Microstructure, that is an ordering on mixed variables, then there exists an ordering for which P satisfies BTP considering its DR-Microstructure.*

3.3   Microstructures and "0-1-all" Constraints

In the previous subsections, it seems that the DR-Microstructure should be the most interesting. Does this feeling remains true for other tractable classes? To begin the study we analyze the well known tractable class defined by *Zero-One-All constraints* ("0-1-all") introduced in [24]. Firstly, we recall the definition:

Definition 10. *A binary CSP $P = (X, D, C)$ is said **0-1-all** (ZOA) if for each constraint C_{ij} of C, for each value $v_i \in D_{x_i}$, C_{ij} satisfies one of the following conditions:*

- *(0) for any value $v_j \in D_{x_j}$, $(v_i, v_j) \notin R(C_{ij})$,*
- *(1) there is a unique value $v_j \in D_{x_j}$ such that $(v_i, v_j) \in R(C_{ij})$,*
- *(all) for any value $v_j \in D_{x_j}$, $(v_i, v_j) \in R(C_{ij})$.*

This property can be represented graphically using the microstructure. In the case of the DR-Microstructure, it seems easy to define the same kind of property. With respect to the case of the definition given above (the one of [24] defined for binary CSPs), the difference will be related to the fact that the edges of the DR-Microstructure connect now tuples of relations. So, since there is no particular feature which can be immediately deduced from the new representation, the satisfaction of the "0-1-all" property is obviously related to the properties of the considered instance.

For the HT-Microstructure, now, the edges connect tuples (vertices associated to constraints of the CSP) to values (vertices associated to variables of the CSP). We now analyze these edges from two viewpoints, i.e. from the two possible directions.

- *Edges from the tuples to the values.* Each tuple is connected to the values appearing in the tuple. So, for each constraint associated to the HT-Microstructure, the connection is a "one" connection, satisfying the conditions of the "0-1-all" property.
- *Edges coming from the values to the tuples.* For a constraint associated to the HT-Microstructure, a value is connected to the tuples where it appears. We discuss the three possibilities.
  - "0" connection. A value is supported by no tuple. For a binary CSP, it is the same case as for the classical definition, with a connection "0".
  - "1" connection. A value is supported by one tuple. For a binary CSP, it is also the same case as for the classical definition, with a connection "1".
  - "all" connection. A value is supported by all the tuples of a constraint. We have also the same configuration as for the "all" connections in the case of binary CSPs.

So, it is quite possible to have instances satisfying the "0-1-all" property for the HT-Microstructure.

Finally, for the ME-Microstructure, we must verify simultaneously the conditions defined for the DR and HT-Microstructures because, the additional edges connecting vertices associated to values correspond to universal constraints, which trivially satisfy the "0-1-all" property.

To conclude, by construction, nothing is opposite to satisfy the conditions of ZOA, even if, as for the case of binary CSPs, these conditions are really restrictive.

4   Conclusion

In this paper, we have introduced the concept of microstructure in the case of CSP with constraints of arbitrary arity. If the concept of microstructure of binary CSP is now well established and has enabled to provide the basis for many theoretical works in CSPs, for the general case, the notion of microstructure was not clearly established before. Also, in this paper, we have wanted to define explicitly a microstructure of CSP for the general case. The idea is to provide a tool for the theoretical study of CSP with constraints of any arity.

Three proposals are presented here: the DR-Microstructure (and the associated DSR-Microstructures), the HT-Microstructure and the ME-Microstructure. Actually, they are derived from the representation of non binary CSPs by equivalent binary CSPs: the dual representation, the hidden variable transformation, and the mixed approach. We have studied these different microstructures whose none constitutes a formal generalization of the classical binary microstructure.

Although this work is prospective, we have begun to show the interest of this approach. For this, we have studied some known tractable classes which have been initially defined for binary CSPs, and expressed in terms of properties of the microstructure of binary CSPs. Here, a first result is related to the case of microstructures of binary CSP whose the number of maximal cliques is bounded by a polynomial. These instances are known to be tractable in polynomial time by the usual algorithms for solving binary CSPs, as BT, FC or RFL. These classes extend naturally to non binary CSPs whose microstructures satisfy the same properties about the number of maximal cliques, if now using the non binary versions of the same algorithms. We have also shown how the BTP class can naturally be extended to non-binary CSPs while expressing the notion of broken triangle within a microstructure of non binary CSPs. This class is of interest because it includes various well-known tractable classes of binary CSPs, which are now defined in terms of constraints of arbitrary arity. We now hope that these tools will be used at the level of non binary CSPs for theoretical studies as it was the case for the classical microstructure of binary CSPs. Although a practical use of these microstructures seems quite difficult for us with respect to issues of efficiency, we believe that one possible and promising track of this work could be to better understand how common backtracking algorithms work efficiently for the non binary case, and the same thing for numerous heuristics.

References

1. Rossi, F., van Beek, P., Walsh, T.: Handbook of Constraint Programming. Elsevier (2006)
2. Freuder, E.: A Sufficient Condition for Backtrack-Free Search. Journal of the ACM 29(1), 24–32 (1982)
3. Beeri, C., Fagin, R., Maier, D., Yannakakis, M.: On the desirability of acyclic database schemes. Journal of the ACM 30, 479–513 (1983)
4. Janssen, P., Jégou, P., Nouguier, B., Vilarem, M.C.: A filtering process for general constraint satisfaction problems: achieving pairwise-consistency using an associated binary representation. In: Proceedings of IEEE Workshop on Tools for Artificial Intelligence, pp. 420–427 (1989)
5. Dechter, R., Pearl, J.: Tree-Clustering for Constraint Networks. Artificial Intelligence 38, 353–366 (1989)
6. Jégou, P., Terrioux, C.: Hybrid backtracking bounded by tree-decomposition of constraint networks. Artificial Intelligence 146, 43–75 (2003)
7. Garey, M.R., Johnson, D.S.: Computer and Intractability. Freeman (1979)
8. Jégou, P.: Decomposition of Domains Based on the Micro-Structure of Finite Constraint Satisfaction Problems. In: Proceedings of AAAI, pp. 731–736 (1993)
9. Gavril, F.: Algorithms for minimum coloring, maximum clique, minimum covering by cliques, and maximum independent set of a chordal graph. SIAM Journal on Computing 1(2), 180–187 (1972)
10. Cohen, D.A.: A New Class of Binary CSPs for which Arc-Consistency Is a Decision Procedure. In: Rossi, F. (ed.) CP 2003. LNCS, vol. 2833, pp. 807–811. Springer, Heidelberg (2003)
11. Salamon, A.Z., Jeavons, P.G.: Perfect Constraints Are Tractable. In: Stuckey, P.J. (ed.) CP 2008. LNCS, vol. 5202, pp. 524–528. Springer, Heidelberg (2008)

12. Cooper, M., Jeavons, P., Salamon, A.: Generalizing constraint satisfaction on trees: hybrid tractability and variable elimination. Artificial Intelligence 174, 570–584 (2010)
13. El Mouelhi, A., Jégou, P., Terrioux, C., Zanuttini, B.: On the Efficiency of Backtracking Algorithms for Binary Constraint Satisfaction Problems. In: ISAIM (January 2012)
14. Angelsmark, O., Jonsson, P.: Improved Algorithms for Counting Solutions in Constraint Satisfaction Problems. In: Rossi, F. (ed.) CP 2003. LNCS, vol. 2833, pp. 81–95. Springer, Heidelberg (2003)
15. Cohen, D.A., Jeavons, P., Jefferson, C., Petrie, K.E., Smith, B.M.: Symmetry definitions for constraint satisfaction problems. Constraints 11(2-3), 115–137 (2006)
16. Mears, C., de la Banda, M.G., Wallace, M.: On implementing symmetry detection. Constraints 14(4), 443–477 (2009)
17. Stergiou, K., Walsh, T.: Encodings of Non-Binary Constraint Satisfaction Problems. In: Proceedings of AAAI, pp. 163–168 (1999)
18. Dechter, R., Pearl, J.: Network-based heuristics for constraint satisfaction problems. Artificial Intelligence 34, 1–38 (1987)
19. Bernstein, P.A., Goodman, N.: The power of natural semijoins. SIAM Journal on Computing 10(4), 751–771 (1981)
20. El Mouelhi, A., Jégou, P., Terrioux, C., Zanuttini, B.: Some New Tractable Classes of CSPs and their Relations with Backtracking Algorithms. In: Gomes, C., Sellmann, M. (eds.) CPAIOR 2013. LNCS, vol. 7874, pp. 61–76. Springer, Heidelberg (2013)
21. Jégou, P.: Contribution à l'étude des problèmes de satisfaction de contraintes: Algorithmes de propagation et de résolution – Propagation de contraintes dans les réseau dynamiques. PhD thesis, Université des Sciences et Techniques du Languedoc (January 1991)
22. Peirce, C.S., Hartshorne, C., Weiss, P.: Collected Papers of Charles Sanders Peirce, vol. 3. Harvard University Press (1933)
23. Rossi, F., Petrie, C.J., Dhar, V.: On the Equivalence of Constraint Satisfaction Problems. In: Proceedings of ECAI, pp. 550–556 (1990)
24. Cooper, M., Cohen, D., Jeavons, P.: Characterising Tractable Constraints. Artificial Intelligence 65(2), 347–361 (1994)
25. Haralick, R., Elliot, G.: Increasing tree search efficiency for constraint satisfaction problems. Artificial Intelligence 14, 263–313 (1980)
26. Nadel, B.: Tree Search and Arc Consistency in Constraint-Satisfaction Algorithms. In: Search in Artificial Intelligence, pp. 287–342. Springer (1988)
27. Dujmovic, V., Fijavz, G., Joret, G., Sulanke, T., Wood, D.R.: On the maximum number of cliques in a graph embedded in a surface. European Journal of Combinatorics 32(8), 1244–1252 (2011)
28. Chmeiss, A., Jégou, P.: A generalization of chordal graphs and the maximum clique problem. Information Processing Letters 62, 111–120 (1997)
29. Graham, M.H.: On the universal relation. Technical report, University of Toronto (1979)
30. Fagin, R.: Degrees of Acyclicity for Hypergraphs and Relational Database Schemes. Journal of the ACM 30(3), 514–550 (1983)
31. El Mouelhi, A., Jégou, P., Terrioux, C.: A Hybrid Tractable Class for Non-Binary CSPs. In: Proceedings of ICTAI (2013)

Finding Maximal Common Sub-parse Thickets for Multi-sentence Search

Boris A.Galitsky[1], Dmitry Ilvovsky[2], Sergei O. Kuznetsov[2], and Fedor Strok[2]

[1] eBay Inc San Jose CA USA
`bgalitsky@hotmail.com`
[2] Higher School of Economics, Moscow Russia
`dilv_ru@yahoo.com`, `skuznetsov@hse.ru`, `fdr.strok@gmail.com`

Abstract. We develop a graph representation and learning technique for parse structures for paragraphs of text. We introduce Parse Thicket (PT) as a set of syntactic parse trees augmented by a number of arcs for inter-sentence word-word relations such as co-reference and taxonomic relations. These arcs are also derived from other sources, including Speech Act and Rhetoric Structure theories. We provide a detailed illustration of how PTs are built from parse trees and generalized as phrases by computing maximal common subgraphs. The proposed approach is subject to evaluation in the product search and recommendation domain, where search queries include multiple sentences. We draw the comparison for search relevance improvement by pair-wise sentence generalization, phrase-level generalization, and generalizations of PTs as graphs.

1   Introduction

Starting from the late seventies, graph-based techniques have been proposed as a powerful tool for pattern representation and classification in structural pattern recognition. After the initial enthusiasm induced by the apparent "smartness" of this data structure, graphs have been practically left unused for a long period of time. Recently, the use of graphs is getting a growing attention from the scientific community. Finding the maximal common subgraphs of graphs is important in many applications, such as bioinformatics [10], chemistry [11, 12, 14], suffix trees for taxonomy creation [1] and text mining [15, 53]. Automated learning of graph taxonomies is important to improve search relevance [8], where a tree representation of a complex question is compared with that of a candidate answer and then matched with a taxonomy tree.

In this study we will represent a paragraph of text as a graph derived from syntactic parse trees of the sentences. Such type of representation will be used to support a complex search where search query includes multiple sentences. Maximal common subgraphs will be used to estimate structural similarity between two portions of text, in particular a multi-sentence search query and an answer.

Modern search engines are not very good at tackling queries including multiple sentences. They either find a very similar document, if it is available, or very dissimilar ones, so that search results are not very useful to the user. This is due to the fact

M. Croitoru et al. (Eds.): GKR 2013, LNAI 8323, pp. 39–57, 2014.

that for multi-sentences queries it is rather hard to learn ranking based on user clicks, since the number of longer queries is practically unlimited. Hence we need a linguistic technology which would rank candidate answers based on structural similarity between the question and the answer. In this study we build a graph-based representation for a paragraph of text so that we can track the structural difference between these paragraphs, taking into account not only parse trees but the discourse as well.

Paragraphs of text as queries appear in the search-based recommendation domains [32, 43, 44] and social search [45]. Recommendation agents track user chats, user postings on blogs and forums, user comments on shopping sites, and suggest web documents and their snippets, relevant to a purchase decisions. To do that, these recommendation agents need to take portions of text, produce a search engine query, run it against a search engine API such as Bing or Yahoo, and filter out the search results which are determined to be irrelevant to a purchase decision. The last step is critical for a sensible functionality of a recommendation agent, and poor relevance would lead to lost trust in the recommendation engine. Hence an accurate assessment of similarity between two portions of text is critical to a successful use of recommendation agents.

Parse trees have become a standard form of representing the syntactic structures of sentences [13, 46, 47, 49]. In this study we will attempt to represent a linguistic structure of a paragraph of text based on parse trees for each sentence of this paragraph. We will refer to the set of parse trees plus a number of arcs for inter-sentence relations between nodes for words as Parse Thicket. A PT is a graph which includes parse trees for each sentence, as well as additional arcs for inter-sentence relationship between parse tree nodes for words.

We will define the operation of generalization of text paragraphs via generalization of respective PTs to assess similarity between them. The use of generalization for similarity assessment is inspired by structured approaches to machine learning [38, 40, 41, 42] versus statistical alternatives where similarity is measured by a distance in feature space [36, 39, 50, 51]. Our intention is to extend the operation of least general generalization (e.g., the antiunification of logical formulas [35, 37]) towards structural representations of paragraph of texts. Hence we will define the operation of generalization on a pair of PT as finding the maximal common sub-thickets and outline two approaches to it:

- Based on generalizing **phrases** from two paragraphs of text
- Based on generalizing **graph representation** for PTs for these paragraphs.

This generalization operation is a base for number of text analysis application such as search, classification, categorization, and content generation [3]. Generalization of text paragraphs is based on the operation of generalization of two sentences, explored in our earlier studies [3, 4, 6]. In addition to learning generalizations of individual sentences, we learn how the links between words in sentences other than syntactic ones can be used to compute similarity between texts. We rely on our formalizations of the theories of textual discourse such as Rhetoric Structure Theory [25] to improve the efficiency of text retrieval.

Whereas machine learning of syntactic parse trees for individual sentences is an established area of research, the contribution of this paper is a structural approach to learning of syntactic information at the level of paragraphs. A number of studies applied machine learning techniques to syntactic parse trees [33], convolution kernels [16] being the most popular approach [22, 52].

2   Parse Thickets and Their Graph Representation

2.1   Introducing Parse Thickets

Is it possible to find more commonalities between texts, treating parse trees at a higher level? For that we need to extend the syntactic relations between the nodes of the syntactic dependency parse trees towards more general text discourse relations.

Which relations can we add to the sum of parse trees to extend the match? Once we have such relations as "the same entity", "sub-entity", "super-entity" and anaphora, we can extend the notion of phrase to be matched between texts. Relations between the nodes of parse trees (which are other than syntactic) can merge phrases from different sentences, or from a single sentence which are not syntactically connected. We will refer to such extended phrases as thicket phrases.

If we have two parse trees P_1 and P_2 of text T_1, and an arc for a relation $r : P_{1i} \to P_{2j}$ between the nodes P_{1i} and P_{2j}, we can now match $..., P_{1i-2}, P_{1i-1}, P_{1i}, P_{2j}, P_{2j+1}, P_{2j+2}, ...$ of T_1 against a chunk of a single sentence of merged chunks of multiple sentences from T_2.

2.2   Finding Similarity between Two Paragraphs of Text

We will compare the following approaches to assessing the similarity of text paragraphs:

- Baseline: bag-of-words approach, which computes the set of common keywords/n-grams and their frequencies.
- Pair-wise matching: we will apply syntactic generalization to each pair of sentences, and sum up the resultant commonalities. This technique has been developed in our previous work [6, 8].
- Paragraph-paragraph matching.

The first approach is most typical for industrial NLP applications today, and the second one was used in our previous studies. The kernel-based approach to parse tree similarities [29], as well as tree sequence kernel [28], being tuned to parse trees of individual sentences, also belongs to the second approach.

We intend to demonstrate the richness of the approach being proposed, and in the consecutive sections we will provide a step-by-step explanation. We will introduce a pair of short texts (articles) and compare the above three approaches. The first paragraph can be viewed as a search query, and the second paragraph can be viewed as a candidate answer. A relevant answer should be a closely related text, which is not a piece of duplicate information.

Note. " ∧ " in the following example and through all the paper means generalization operation. Describing parse trees we use standard notation for constituency trees: [...] represents sub-phrase, NN, JJ, NP etc. denote parts-of-speech and phrases (noun, adjective, noun phrase resp.), * is used to denote random tree node.

"Iran refuses to accept the UN proposal to end the dispute over work on nuclear weapons",

"UN nuclear watchdog passes a resolution condemning Iran for developing a second uranium enrichment site in secret",

"A recent IAEA report presented diagrams that suggested Iran was secretly working on nuclear weapons",

"Iran envoy says its nuclear development is for peaceful purpose, and the material evidence against it has been fabricated by the US",

∧

"UN passes a resolution condemning the work of Iran on nuclear weapons, in spite of Iran claims that its nuclear research is for peaceful purpose",

"Envoy of Iran to IAEA proceeds with the dispute over its nuclear program and develops an enrichment site in secret",

"Iran confirms that the evidence of its nuclear weapons program is fabricated by the US and proceeds with the second uranium enrichment site"

1. The list of **common keywords** gives a hint that both documents are on nuclear program of Iran, however it is hard to get more specific details.

Iran, UN, proposal, dispute, nuclear, weapons, passes, resolution, developing, enrichment, site, secret, condemning, second, uranium

2. **Pair-wise generalization** gives a more accurate account on what is common between these texts.

[NN-work IN- IN-on JJ-nuclear NNS-weapons],   [DT-the NN-dispute IN-over JJ-nuclear NNS-*], [VBZ-passes DT-a NN-resolution],*
[VBG-condemning NNP-iran IN-],*
[VBG-developing DT- NN-enrichment NN-site IN-in NN-secret]],*
[DT- JJ-second NN-uranium NN-enrichment NN-site]],*
[VBZ-is IN-for JJ-peaceful NN-purpose],
[DT-the NN-evidence IN- PRP-it],   [VBN-* VBN-fabricated IN-by DT-the NNP-us]*

3. **Parse Thicket generalization** gives the detailed similarity picture which looks more complete than the pair-wise sentence generalization result above:

[NN-Iran VBG-developing DT- NN-enrichment NN-site IN-in NN-secret]*

[NN-generalization-<UN/nuclear watchdog> * VB-pass NN-resolution VBG condemning NN- Iran]

[NN-generalization-<Iran/envoy of Iran> Communicative_action   DT-the NN-
dispute IN-over JJ-nuclear NNS-*

[Communicative_action - NN-work   IN-of NN-Iran IN-on JJ-nuclear NNS-
weapons]

[NN-generalization <Iran/envoy to UN> Communicative_action   NN-Iran NN-
nuclear NN-* VBZ-is IN-for JJ-peaceful NN-purpose],

Communicative_action - NN-generalize <work/develop> IN-of NN-Iran IN-on JJ-
nuclear NNS-weapons]*

[NN-generalization <Iran/envoy to UN> Communicative_action NN-evidence IN-
against NN Iran NN-nuclear   VBN-fabricated IN-by DT-the NNP-us]

condemn^proceed [enrichment site] <leads to>   suggest^condemn [work Iran
nuclear weapon]

One can feel that PT-based generalization closely approaches human performance
in terms of finding similarities between texts. To obtain these results, we need to be
capable of maintaining coreferences, apply the relationships between entities to our
analysis (subject vs relation-to-this subject), including relationships between verbs
(develop is a partial case of work). We also need to be able to identify communicative
actions and generalize them together with their subjects according to the specific patterns of speech act theory. Moreover, we need to maintain rhetoric structure relationship between sentences, to generalize at a higher level above sentences.

The focus of this paper will be to introduce parse thicket and their generalization as
a paragraph-level structured representation. It will be done with the help of the above
example.

2.3   Arcs of Parse Thicket Based on Theories of Discourse

We attempt to treat computationally, with a unified framework, two approaches to
textual discourse:

- Rhetoric structure theory (RST) [25];
- Speech Act theory (SpActT) [26].

Although both these theories have psychological observation as foundations and are
mostly of a non-computational nature, we will build a specific computational
framework for them [4]. We use these sources to find links between sentences to
enhance indexing for search. For RST, we attempt to extract an RST relation and
form a thicket phrase around it, including a placeholder for RST relation itself. For
SpActT, we use a vocabulary of communicative actions to find their subjects [5], add
respective arcs to PT and form the respective set of thicket phrases.

RST Example

Fig. 1 shows the generalization of two RST relations "RST-evidence". This pair of
relations occurs between phrases evidence-for-what   [Iran's nuclear weapon
program] and what-happens-with-evidence [Fabricated by USA] in PT_1 and
evidence-for-what [against Iran's nuclear development] and what-happens-with-
evidence [Fabricated by the USA] in PT_2.

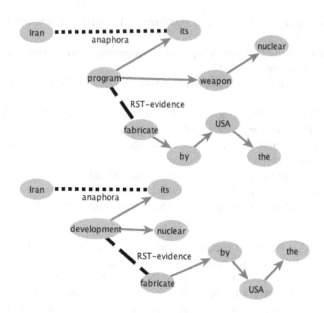

Fig. 1. Two fragments of two PTs showing the generalization for the pairs of RST relations

Notice that in the latter case we need to merge (perform anaphora substitution) the phrase '*its nuclear development*' with '*evidence against it*' to obtain '*evidence against its nuclear development*'. Notice the arc *it - development*, according to which this anaphora substitution occurred. Evidence is removed from the phrase because it is the indicator of RST relation, and we form the subject of this relation to match. Furthermore, we need another anaphora substitution *its - Iran* to obtain the final phrase.

As a result of generalizations of two RST relations of the same sort (evidence) we obtain *Iran nuclear NNP – RST-evidence – fabricate by USA*. Notice that we could not obtain this similarity expression by using sentence-level generalization.

Communicative Actions Example

Communicative actions are used by text authors to indicate a structure of a dialogue or a conflict [26]. Hence analyzing the communicative actions' arcs of PT, one can find implicit similarities between texts. We can generalize:

1. one communicative actions with its subject from T_1 against another communicative action with its subject from T_2 (communicative action arc is not used) ;

2. a pair of communicative actions with their subjects from T_1 against another pair of communicative actions from T_2 (communicative action arcs are used).

In our example (Figs. 2 & 3) we have the same communicative actions with subjects with low similarity: *condemn ['Iran for developing second enrichment site in secret']* *vs condemn ['the work of Iran on nuclear weapon']* or different communicative actions with similar subjects.

The two distinct communicative actions dispute and condemn have rather similar subjects: 'work on nuclear weapon'. Generalizing two communicative actions with their subjects follows the rule: generalize communicative actions themselves, and 'attach' the result to generalization of their subjects as regular sub-tree generalization. Two communicative actions can always be generalized, which is not the case for their subjects: if their generalization result is empty, the generalization result of communicative actions with these subjects is empty too. The generalization result here for the case 1 above is: *condemn^dispute [work-Iran-on-nuclear-weapon].*

Generalizing two different communicative actions is based on their attributes and is presented elsewhere [4, 5].

T_1 T_2

condemn [second uranium enrichment site] ↔ proceed [develop an enrichment site in secret] ↓   *communicative action arcs*   ↓ suggest [Iran is secretly working on nuclear weapon] ↔ condemn [the work of Iran on nuclear weapon]

results in *condemn^proceed [enrichment site] <leads to>   suggest^condemn [work Iran nuclear weapon].*

Notice that generalization

condemn [second uranium enrichment site] ↔ condemn [the work of Iran on nuclear weapon] ↓   *communicative action arcs*   ↓ suggest [Iran is secretly working on nuclear weapon] ↔ proceed [develop an enrichment site in secret]

gives zero result because the arguments of condemn from T_1 and T_2 are not very similar. Hence we generalize the subjects of communicative actions first before we generalize communicative actions themselves.

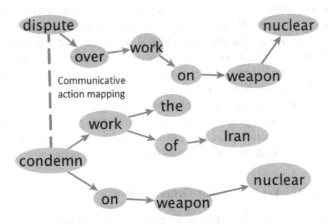

Fig. 2. An example of the mapping for the pair of communicative actions

Fig. 3. Two fragments of two PTs showing the generalization for the pairs of communicative actions

2.4 Phrase-Level Generalization

Although the generalization is defined as the set of maximal common sub-graphs, its computation is based on matching phrases. To generalize a pair of sentences, we perform chunking and extract all noun, verb, prepositional and other types of phrases from each sentence. Then we perform generalization for each type of phrases, attempting to find a maximal common sub-phrase for each pair of phrases of the same type. The resultant phrase-level generalization can then be interpreted as a set of paths in resultant common sub-trees [3].

Generalization of parse thickets, being a maximal common sub-graph (sub-parse thicket) can be computed at the level of phrases as well as a structure containing a maximal common sub-phrases. However, the notion of phrases is extended now: thicket phrases can contain regular phrases from different sentences. The way these phrases are extracted and formed depends on the source of non-syntactic link between words in different sentences: thicket phrases are formed in a different way for communicative actions and RST relations. Notice that the set of regular phrases for a parse thicket is a sub-set of the set of thicket phrases (all phrases extracted for generalization). Because of this richer set of phrases for generalization, the parse thicket generalization is richer than the pair-wise thicket generalization, and can better tackle variety in phrasings and writing styles, as well as distribution of information through sentences.

2.5 Forming Thicket Phrases for Generalization

We will now outline the algorithm of forming thicket phrases. Most categories of thicket arcs will be illustrated below. Please refer to [7] for further details.

For each sentence S in a paragraph P:

1. Form a list of previous sentences in a paragraph S_{prev}
2. For each word in the current sentence:
   (a) If this word is a *pronoun*: find all nouns or noun phrases in the S_{prev} which are:
      o The same entities (via anaphora resolution)

      (i) If this word is a *noun*: find all nouns or noun phrases in the S_{prev} which are:

      o The same entities (via anaphora resolution)
      o Synonymous entity
      o Super entities
      o Sub and sibling entities

   (b) If this word is a *verb*:
      (i) If it is a communicative action:
         (1) Form the phrase for its subject $VBCA_{phrase}$, including its verb phrase VB_{phrase}
         (2) Find a preceding communicative action $VBCA_{phrase0}$ from S_{prev} with its subject
         (3) Form a thicket phrase $[VBCA_{phrase} , VBCA_{phrase0}]$
      (ii) If it indicates *RST relation*:
         (1) Form the phrase for the pair of phrases which are the subjects $[VBRST_{phrase1}, VBRST_{phrase2}]$, of this RST relation, $VBRST_{phrase1}$ belongs to S_{prev}.

Notice the four categories of the used phrases:

- Regular phrases;
- Thicket phrases (entity, sub-entity, coreferences);
- RST phrases;
- SpActT phrases.

2.6 Sentence-Level Generalization Algorithm

Below we outline the main steps of the algorithm for finding a maximal sub-phrase of a pair of phrases, applied to the sets of thicket phrases for T_1 and T_2. Please refer to [3] for further details.

1. Split parse trees for sentences into sub-trees which are phrases for each type: *verb, noun, prepositional* and others; these sub-trees are overlapping. The sub-trees are coded so that the information about occurrence in the full tree is retained.
2. All sub-trees are grouped by phrase types.
3. Extending the list of phrases by adding equivalence transformations
4. Generalize each pair of sub-trees for both sentences for each phrase type.
5. For each pair of sub-trees yield an alignment, and then generalize each node for this alignment. For the obtained set of trees (generalization results), calculate the score.

6. For each pair of sub-trees for phrases, select the set of generalizations with the highest score (the least general).
7. Form the sets of generalizations for each phrase types whose elements are sets of generalizations for this type.
8. Filtering the list of generalization results: for the list of generalization for each phrase type, exclude more general elements from lists of generalization for given pair of phrases.

3   Computing Maximal Common Sub-PTs

To find maximal subPT we use a reduction of the maximal common subgraph problem to the maximal clique problem [27] by constructing the edge product. The main difference with the traditional edge product is that instead of requiring the same label for two edges to be combined, we require non-empty generalization results for these edges. Though computationally not optimal, this approach is convenient for prototyping.

To solve the maximal subPT problem, we transform it to the maximal clique problem [27] . This can be done by constructing an edge product PT. Let $G_1 = (V_1, E_1, \alpha_1, L_1)$ and $G_2 = (V_2, E_2, \alpha_{21}, L_2)$ be PTs with nodes V and edges E, where $\alpha : V \to L$ is a function assigning labels to the vertices, and L is a finite non-empty set of labels for the edges and vertices. The edge product PT $H_e = G_1 \circ G_2$ includes the vertex set $V_H = E_1 \circ E_2$ in which all edge pairs (e_i, e_j) with $1 \le i \le |E_1|$ and $1 \le j \le |E_2|$ have to produce non-empty generalization in their edge labels. Also, these edge pairs must produce non-empty generalizations in their corresponding end vertex labels. Let $e_i = (u_1, v_1, l_1)$ and $e_j = (u_2, v_2, l_2)$. The labels coincide if $l_1 \cap l_2 \ne \varnothing$ and $\alpha_1(v_1) \cap \alpha_2(v_2) \ne \varnothing$ and $\alpha_1(u_1) \cap \alpha_2(u_2) \ne \varnothing$.

There is an edge between vertices $e_H, f_H \in V_H$ where $e_H = (e_1, e_2)$ and $f_H = (f_1, f_2)$ if two edge pairs are compatible, meaning that they are distinct: $e_1 \ne f_1$ and $e_2 \ne f_2$. Also, either condition for these edges of resultant PT should hold:

- e_1, f_1 in G_1 are connected via a vertex of the label which produces non-empty generalization with the label of the vertex shared by e_2, f_2 in G_2: $\{vertices\ for\ e_1, f_1\} \cap \{vertices\ for\ e_2, f_2\} \ne \varnothing$. We label and call them *c-edges*, or
- e_1, f_1 and e_2, f_2 are not adjacent in G_1 and in G_2, respectively. We label and call them *d-edges*.

To get a common subPT in G_1 and G_2 each edge pair in G_1 and G_2 (vertex pair in H_e) has to have generalizable label with all other edge pairs in G_1 and G_2 (vertex pair in H_e), which are forming a common subgraph. In this way a clique in H_e corresponds to common subPTs G_1 and G_2.

A subset of vertices of G is said to be a clique if all its nodes are mutually adjacent. A maximal clique is one which is not contained in any larger clique, while a maximum clique is a clique having largest cardinality.

After finding all **maximal** cliques for all pairs (representing question and answer for instance) we look through all results and range them due to cliques cardinality. In this case the more edge pairs the result of generalization of two paragraphs contains the more relevant it is.

Applying some effective (taking into account specific features of thicket graphs) approaches to common subgraph problem is a subject of our future work. Also we are going to use more complex types of generalization such as pattern structures [9] instead of pair-wise generalization.

4   Architecture of PT Processing System

There are three system components, which include Parse Thicket building, phrase-level processing and graph processing (Fig. 4).

The textual input is subject to a conventional text processing flow such as sentence splitting, tokenizing, stemming, part-of-speech assignment, building of parse trees and coreferences assignment for each sentence. This flow is implemented by either OpenNLP or Stanford NLP, and the parse thicket is built based on the algorithm presented in this paper. The coreferences and RST component strongly relies on Stanford NLP's rule-based approach to finding correlated mentions, based on the multi-pass sieves.

The graph-based approach to generalization relies on finding maximal cliques for an edge product of the graphs for PTs being generalized. As it was noticed earlier the main difference with the traditional edge product is that instead of requiring the same label for two edges to be combined, we require non-empty generalization results for these edges. Hence although the parse trees are relatively simple graphs, parse thicket graphs reach the limit of real-times processing by graph algorithms.

The system architecture serves as a basis of OpenNLP – similarity component, which is a separate Apache Software foundation project, accepting input from either OpenNLP or Stanford NLP. It converts parse thicket into JGraphT objects which can be further processed by an extensive set of graph algorithms. In particular, finding maximal cliques is based on [27] algorithm. Code and libraries described here are also available at http://code.google.com/p/relevance-based-on-parse-trees and http://svn.apache.org/repos/asf/opennlp/sandbox/opennlp-similarity/. The system is ready to be plugged into Lucene library to improve search relevance. Also, a SOLR request handler is provided so that search engineers can switch to a PT-based multi-sentence search to quickly verify if relevance is improved.

The second and third components are two different ways to assess similarity between two parse thickets. Phrase-level processing for the phrases of individual sentences has been described in detail in our previous studies [3,5]. In this study we collect all phrases for all sentences of one paragraph of text, augment them with thicket phrases (linguistic phrases which are merged based on the inter-sentence relation), and generalize against that of the other paragraph of text.

5   Algorithms and Scalability of the Approach

The generalization operation on parse trees for sentences and parse thickets for paragraphs is defined as finding a set of maximum common sub-trees and sub parse thickets respectively. Although for the trees this problem is $O(n)$, for the general case of graphs finding maximal common sub-graphs is NP hard [17].

One approach to learning parse trees is based on tree kernels, where the authors propose a technique oriented specifically to parse trees, and reduce the space of all possible sub-trees. Partial tree kernel [22] allows partial rule matching by ignoring some child nodes in the original production rule. Tree Sequence Kernel or TSK [28] adopts the structure of a sequence of sub-trees other than the single tree structure, strictly complying with the original production rules. Leveraging sequence kernel and tree kernel, TSK enriches sequence kernel with syntactic structure information and enriches tree kernel with disconnected sub-tree sequence structures.

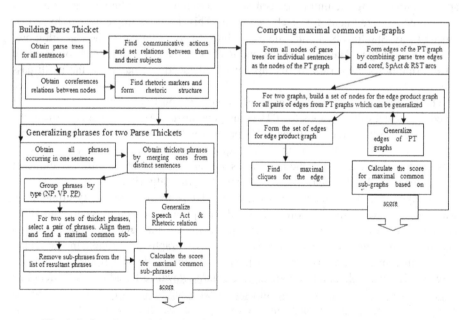

Fig. 4. A chart for graph-based and phrase-based generalization of parse thickets

The approach of phrase-level matching for parse trees and thicket phrase matching for parse thickets turns out to be much more efficient than graph based approaches, including graph kernels. Instead of considering the space of all possible sub-graphs, we consider paths in trees and graphs, which correspond to linguistic phrases, or to phrases merged according to inter-sentence relations of the discourse. We do not consider parts of trees or thickets which correspond neither to a phrase of parse tree nor to a thicket phrase of thicket. This is due to the observation that such sub-trees and sub-graphs are incomplete, redundant or noisy features to rely on, conducting learning.

To estimate the complexity of generalization of two parse thickets, let us consider an average case with five sentences in each paragraph and 15 words in each sentence. Such thickets have on average 10 phrases per sentence, 10 inter-sentence arcs, which give us up to 40 thicket phrases each. Hence for such parse thickets we have to generalize up to 50 linguistic phrases and 40 thicket phrases of the first thicket against the set of similar size for the second thicket. Taking into account a separate generalization of noun and verb phrases, this average case consists of $2* 45*45$ generalizations, followed by the subsumption checks. Each phrase generalization is based on up to 12 string comparisons, taking an average size of phrase as 5 words. Hence on average the parse thicket generalization includes $2*45*45*12*5$ operations. Since a string comparison takes a few microseconds, thicket generalization takes on average 100 milliseconds without use of index. However, in an industrial search application where phrases are stored in an inverse index, the generalization operation can be completed in constant time, irrespectively of the size of index [18]. In case of map-reduce implementation of generalization operation, for example, using Cascading framework, the time complexity becomes constant with the size of candidate search results to be re-ranked [20].

6   Evaluation of Multi-sentence Search

Having described the system architecture and engineering aspects, we proceed to evaluation of how generalization of PTs can improve multi-sentence search, where one needs to compare a query as a paragraph of text against a candidate answer as a paragraph of text (snippet). We conducted evaluation of relevance of syntactic generalization – enabled search engine, based on Yahoo and Bing search engine APIs. For an individual query, the relevance was estimated as a percentage of correct hits among the first thirty, using the values: {correct, marginally correct, incorrect}. Accuracy of a single search session is calculated as the percentage of correct search results plus half of the percentage of marginally correct search results. Accuracy of a particular search setting (query type and search engine type) is calculated, averaging through 100 search sessions. For our evaluation, we use customers' queries to eBay entertainment and product-related domains, from simple questions referring to a particular product, a particular user need, as well as a multi-sentence forum-style request to share a recommendation. In our evaluation we split the totality of queries into noun-phrase class, verb-phrase class, how-to class, and also independently split in accordance to query length (from 3 keywords to multiple sentences). The evaluation was conducted by the authors.

To compare the relevance values between search settings, we used first 30 search results and re-ranked them according to the score of the given search setting. The list of products which serves the basis of queries is available at code.google.com/p/relevance-based-on-parse-trees/downloads/detail?name=Queries900set.xls. We took each product and found a posting somewhere on the web (typically, a blog or forum posting) about this product, requesting a particular information or addressing a particular user feature, need or concern. From such extended expression containing product names, we formed the list queries of desired complexity. We refer the reader to [3] for the further details on evaluation settings.

Table 1. Evaluation results

Query type	Query complexity	Relevance of baseline Bing search, %, averaging over 100 searches	Relevance single-sentence phrase-based generalization search, %, averaging over 100 searches	Relevance of thicket-based phrase generalization search, %, averaging over 100 searches	Relevance of parse thicket-based graph generalization search, %, averaging over 100 searches
Product recommendation search	1 compound sentence	62.3	69.1	72.4	73.3
	2 sent	61.5	70.5	71.9	71.6
	3 sent	59.9	66.2	72	71.4
	4 sent	60.4	66	68.5	66.7
Travel recommendation search	1 compound sent	64.8	68	72.6	74.2
	2 sent	60.6	65.8	73.1	73.5
	3 sent	62.3	66.1	70.9	72.9
	4 sent	58.7	65.9	72.5	71.7
Facebook friend agent support search	1 compound sent	54.5	63.2	65.3	67.2
	2 sent	52.3	60.9	62.1	63.9
	3 sent	49.7	57	61.7	61.9
	4 sent	50.9	58.3	62	62.7
Average		**58.15**	**64.75**	**68.75**	**69.25**

Evaluation results are shown in Table 1. Three domains are used in evaluation:

- Product recommendation, where an agent reads chats about products and finds relevant information on the web about a particular product.
- Travel recommendation, where an agent reads chats about travel and finds relevant information on the travel websites about a hotel or an activity.
- Facebook recommendation, where an agent reads wall postings and chats, and finds a piece of relevant information for friends on the web.

In each of these domains we selected a portion of text on the web to form a query, and then filtered search results delivered by Bing search engine API. One can observe that unfiltered precision is 58.2%, whereas improvement by pair-wise sentence generalization is 6%, thicket phrases – additional 4%, and graphs – additional 0.5%. One can also see that the higher the complexity of sentence, the higher the contribution of generalization technology, from sentence level to thicket phrases to graphs.

7   Conclusions

In our previous works we observed how employing a richer set of linguistic information such as syntactic relations between words assists relevance tasks [2-8]. To take advantage of semantic discourse information, we introduced parse thicket representation and proposed the way to compute similarity between texts based on generalization of parse thickets. In this work we build the framework for generalizing PTs as sets of phrases on one hand, and generalizing PTs as graphs via maximal common subgraphs, on the other hand.

The operation of generalization to learn from parse trees for a pair of sentences turned out to be important for text relevance tasks. Once we extended it to learning parse thickets for two paragraphs, we observed that the relevance is further increased compared to the baseline (Bing search engine API), which relies on keyword statistics in the case of multi-sentence query. Parse thicket is intended to represent the syntactic structure of text as well as a number of semantic relations for the purpose of indexing for search. To accomplish this, parse thicket includes relations between words in different sentences, such that these relations are essential to match queries with portions of texts to serve as answers.

We considered the following sources of relations between words in sentences: co-references, taxonomic relations such as sub-entity, partial case, predicate for subject etc., rhetoric structure relation and speech acts. We demonstrated that search relevance can be improved if search results are subject to confirmation by parse thicket generalization, when answers occur in multiple sentences.

Traditionally, machine learning of linguistic structures is limited to keyword forms and frequencies. At the same time, most theories of discourse are not computational, they model a particular set of relations between consecutive states. In this work we attempted to achieve the best of both worlds: learn complete parse tree information augmented with an adjustment of discourse theory allowing computational treatment.

We believe this is one of the first studies in learning a semantic discourse to solve a search relevance problem, based on a set of parse trees. Instead of using linguistic information of individual sentences, we can now compute text similarity at the level of paragraphs. We have contributed the PT-based functionality to OpenNLP so that search engineers can easily plug it in their search infrastructure.

We mentioned that the representation of syntactic parse trees based on tree kernels has been extensively exploited in the machine learning tasks. It would be interesting as a future work to use a kernel-based approach in combination with parse thicket structure. We will address the issue of how kernel-based approaches would benefit from the richer parse thicket structure used in this paper.

Another interesting thing to do is to improve our graph processing. Possible solution is related to applying a projections technique [9]. The graph is represented as a set of substructures having exact type (for instance, chains, subtrees, etc.). The operation of finding maximal common substructure for such type of structures (phrases, trees, parse thickets) could be done much faster than under a straight-forward graph matching approach. We will also enrich corpus-based learning of rhetoric structures.

References

1. [Chernyak and Mirkin 2013] Chernyak, E.L., Mirkin, B.G.: Computationally refining a taxonomy by using annotated suffix trees over Wikipedia resources. In: International Conference "Dialogue", RGGU, vol. 12(19) (2013)
2. [Galitsky, 2003] Galitsky, B.: Natural Language Question Answering System: Technique of Semantic Headers. In: Advanced Knowledge International, Australia (2003)
3. [Galitsky et al., 2012] Galitsky, B., de la Rosa, J.L., Dobrocsi, G.: Inferring the semantic properties of sentences by mining syntactic parse trees. Data & Knowledge Engineering 81-82, 21–45 (2012)
4. [Galitsky et al., 2013a] Galitsky, B.A., Kuznetsov, S.O., Usikov, D.: Parse Thicket Representation for Multi-sentence Search. In: Pfeiffer, H.D., Ignatov, D.I., Poelmans, J., Gadiraju, N. (eds.) ICCS 2013. LNCS, vol. 7735, pp. 153–172. Springer, Heidelberg (2013)
5. [Galitsky and Kuznetsov, 2008] Galitsky, B., Kuznetsov, S.: Learning communicative actions of conflicting human agents. J. Exp. Theor. Artif. Intell. 20(4), 277–317 (2008)
6. [Galitsky, 2012] Galitsky, B.: Machine Learning of Syntactic Parse Trees for Search and Classification of Text. In: Engineering Application of AI (2012), http://dx.doi.org/10.1016/j.engappai.2012.09.017
7. [Galitsky et al., 2013b] Galitsky, B., Ilvovsky, D., Kuznetsov, S.O., Strok, F.: Text Retrieval Efficiency with Pattern Structures on Parse Thickets. In: Workshop "Formal Concept Analysis Meets Information Retrieval" at ECIR 2013, Moscow, Russia (2013)
8. [Galitsky 2013] Galitsky, B.: Transfer learning of syntactic structures for building taxonomies for search engines. Engineering Application of AI, http://dx.doi.org/10.1016/j.engappai.2013.08.010
9. [Ganter and Kuznetsov 2001] Ganter, B., Kuznetsov, S.O.: Pattern Structures and Their Projections. In: Delugach, H.S., Stumme, G. (eds.) ICCS 2001. LNCS (LNAI), vol. 2120, pp. 129–142. Springer, Heidelberg (2001)
10. [Ehrlich and Rarey, 2011] Ehrlich, H.-C., Rarey, M.: Maximum common subgraph isomorphism algorithms and their applications in molecular science: review. Wiley Interdisciplinary Reviews: Computational Molecular Science 1(1), 68–79 (2011)

11. [Yan and Han, 2002] Yan, X., Han, J.: gSpan: Graph-Based Substructure Pattern Mining. In: Proc. IEEE Int. Conf. on Data Mining, ICDM 2002, pp. 721–724. IEEE Computer Society (2002)
12. [Blinova et al., 2003] Blinova, V.G., Dobrynin, D.A., Finn, V.K., Kuznetsov, S.O., Pankratova, E.S.: Toxicology Analysis by Means of the JSM-Method. Bioinformatics 19, 1201–1207 (2003)
13. [Punyakanok et al., 2004] Punyakanok, V., Roth, D., Yih, W.: Mapping dependencies trees: an application to question answering. In: Proceedings of AI & Math., Florida, USA (2004)
14. [Kuznetsov and Samokhin, 2005] Kuznetsov, S.O., Samokhin, M.V.: Learning Closed Sets of Labeled Graphs for Chemical Applications. In: Kramer, S., Pfahringer, B. (eds.) ILP 2005. LNCS (LNAI), vol. 3625, pp. 190–208. Springer, Heidelberg (2005)
15. [Wu et al., 2011] Wu, J., Xuan, Z., Pan, D.: Enhancing text representation for classification tasks with semantic graph structures. International Journal of Innovative Computing, Information and Control (ICIC) 7(5(B))
16. [Haussler, 1999] Haussler, D.: Convolution kernels on discrete structures (1999)
17. [Kann, 1992] Kann, V.: On the Approximability of the Maximum Common Subgraph Problem. In: Finkel, A., Jantzen, M. (eds.) STACS 1992. LNCS, vol. 577, pp. 377–388. Springer, Heidelberg (1992)
18. [Lin, 2013] Lin, J.: Data-Intensive Text Processing with MapReduce (2013), http://intool.github.io/MapReduceAlgorithms/MapReduce-book-final.pdf
19. Cascading (2013), http://en.wikipedia.org/wiki/Cascading, http://www.cascading.org/
20. [Dean, 2009] Dean, J.: Challenges in Building Large-Scale Information Retrieval Systems (2009), http://research.google.com/people/jeff/WSDM09-keynote.pdf
21. [Widlöcher 2012] Widlöcher, A., Mathet, Y.: The Glozz platform: a corpus annotation and mining tool. In: ACM Symposium on Document Engineering, pp. 171–180 (2012)
22. [Moschitti, 2006] Moschitti, A.: Efficient Convolution Kernels for Dependency and Constituent Syntactic Trees. In: Fürnkranz, J., Scheffer, T., Spiliopoulou, M. (eds.) ECML 2006. LNCS (LNAI), vol. 4212, pp. 318–329. Springer, Heidelberg (2006)
23. [Mihalcea and Tarau, 2004] Mihalcea, R., Tarau, P.: TextRank: Bringing Order into Texts. Empirial Methods in NLP 2004 (2004)
24. [Polovina and Heaton, 1992] Polovina, S., Heaton, J.: An Introduction to Conceptual Graphs. AI Expert, 36–43 (1992)
25. [Mann et al., 1992] Mann, W.C., Matthiessen, C.M.I.M., Thompson, S.A.: Rhetorical Structure Theory and Text Analysis. In: Mann, W.C., Thompson, S.A. (eds.) Discourse Description: Diverse Linguistic Analyses of a Fund-raising Text, pp. 39–78. John Benjamins, Amsterdam (1992)
26. [Searle, 1969] Searle, J.: Speech acts: An essay in the philosophy of language. Cambridge University, Cambridge (1969)
27. [Bron and Kerbosch, 1973] Bron, C., Kerbosch, J.: Algorithm 457: finding all cliques of an undirected graph. Commun. ACM (ACM) 16(9), 575–577 (1973)
28. [Sun et al., 2011] Sun, J., Zhang, M., Tan, C.L.: Tree Sequence Kernel for Natural Language. AAAI-25 (2011)
29. [Zhang et al., 2008] Zhang, M., Che, W., Zhou, G., Aw, A., Tan, C., Liu, T., Li, S.: Semantic role labeling using a grammar-driven convolution tree kernel. IEEE Transactions on Audio, Speech, and Language Processing 16(7), 1315–1329 (2008)

30. [Vismara and Benoît, 2008] Vismara, P., Valery, B.: Finding Maximum Common Connected Subgraphs Using Clique Detection or Constraint Satisfaction Algorithms. In: Thi, H.A.L., Bouvry, P., Dinh, T.P. (eds.) MCO 2008. CCIS, vol. 14, pp. 358–368. Springer, Heidelberg (2008)
31. [D. Conte et al., 2004] Conte, D., Foggia, P., Sansone, C., Vento, M.: Thirty years of graph matching in pattern recognition. International Journal of Pattern Recognition and Artificial Intelligence 18(3), 265–298 (2004)
32. [Montaner et al., 2003] Montaner, M., Lopez, B., de la Rosa, J.L.: A Taxonomy of Recommender Agents on the Internet. Artificial Intelligence Review 19(4), 285–330 (2003)
33. [Collins and Duffy, 2002] Collins, M., Duffy, N.: Convolution kernels for natural language. In: Proceedings of NIPS, pp. 625–632 (2002)
34. [Lee et al., 2013] Lee, H., Chang, A., Peirsman, Y., Chambers, N., Surdeanu, M., Jurafsky, D.: Deterministic coreference resolution based on entity-centric, precision-ranked rules. Computational Linguistics 39(4) (2013)
35. [Plotkin, 1970] Plotkin, G.D.: A note on inductive generalization. In: Meltzer, B., Michie, D. (eds.) Machine Intelligence, vol. 5, pp. 153–163. Elsevier North-Holland, New York (1970)
36. [Jurafsky and Martin, 2008] Jurafsky, D., Martin, J.H.: Speech and Language Processing. An Introduction to Natural Language Processing. Computational Linguistics, and Speech Recognition (2008)
37. [Robinson, 1965] Robinson, J.A.: A machine-oriented logic based on the resolution principle. Journal of the Association for Computing Machinery 12, 23–41 (1965)
38. [Mill, 1843] Mill, J.S.: A system of logic, ratiocinative and inductive, London (1843)
39. [Fukunaga, 1990] Fukunaga, K.: Introduction to statistical pattern recognition, 2nd edn. Academic Press Professional, Inc., San Diego (1990)
40. [Finn, 1999] Finn, V.K.: On the synthesis of cognitive procedures and the problem of induction. NTI Series 2, N1-2, pp. 8–45 (1999)
41. [Mitchell, 1997] Mitchell, T.: Machine Learning. McGraw Hill (1997)
42. [Furukawa, 1997] Furukawa, K.: From Deduction to Induction: Logical Perspective. In: Apt, K.R., Marek, V.W., Truszczynski, M., Warren, D.S. (eds.) The Logic Programming Paradigm. Springer (1998)
43. [Bhasker and Srikumar, 2012] Bhasker, B., Srikumar, K.: Recommender Systems in E-Commerce. CUP (2012) ISBN 978-0-07-068067-8
44. [Thorsten et al., 2012] Hennig-Thurau, H., Marchand, A., Marx, P.: Can Automated Group Recommender Systems Help Consumers Make Better Choices? Journal of Marketing 76(5), 89–109 (2012)
45. [Trias et al., 2012] Trias i Mansilla, A., de la Rosa i Esteva, J.L.: Asknext: An Agent Protocol for Social Search. Information Sciences 190, 144–161 (2012)
46. [Punyakanok et al., 2005] Punyakanok, V., Roth, D., Yih, W.: The Necessity of Syntactic Parsing for Semantic Role Labeling. In: IJCAI 2005 (2005)
47. [Domingos and Poon, 2009] Domingos, P., Poon, H.: Unsupervised Semantic Parsing. In: Proceedings of the 2009 Conference on Empirical Methods in Natural Language Processing. ACL, Singapore (2009)
48. [Marcu, 1997] Marcu, D.: From Discourse Structures to Text Summaries. In: Mani, I., Maybury, M. (eds.) Proceedings of ACL Workshop on Intelligent Scalable Text Summarization, Madrid, Spain, pp. 82–88 (1997)
49. [Abney, 1991] Abney, S.: Parsing by Chunks. In: Principle-Based Parsing, pp. 257–278. Kluwer Academic Publishers (1991)

50. [Byun and Lee, 2002] Byun, H., Lee, S.-W.: Applications of Support Vector Machines for Pattern Recognition: A Survey. In: Lee, S.-W., Verri, A. (eds.) SVM 2002. LNCS, vol. 2388, pp. 213–236. Springer, Heidelberg (2002)
51. [Manning and Schütze, 1999] Manning, C., Schütze, H.: Foundations of Statistical Natural Language Processing. MIT Press, Cambridge (1999)
52. [Sun et al., 2011] Sun, J., Zhang, M., Tan, C.: Exploring syntactic structural features for sub-tree alignment using bilingual tree kernels. In: Proceedings of ACL, pp. 306–315 (2010)
53. [Kivimäki et al 2013] Kivimäki, I., Panchenko, A., Dessy, A., Verdegem, D., Francq, P., Bersini, H., Saerens, M.: A Graph-Based Approach to Skill Extraction from Text. In: TextGraphs-8, Graph-based Methods for Natural Language Processing. Workshop at EMNLP 2013, Seattle, USA, October 18 (2013)
54. [Widlöcher and Mathet 2012] Widlöcher, A., Mathet, Y.: The Glozz platform: a corpus annotation and mining tool. In: ACM Symposium on Document Engineering, pp. 171–180 (2012)

Intrusion Detection with Hypergraph-Based Attack Models

Antonella Guzzo, Andrea Pugliese,
Antonino Rullo, and Domenico Saccà

University of Calabria, Italy
{guzzo,apugliese,nrullo,sacca}@dimes.unical.it

Abstract. In numerous security scenarios, given a sequence of logged activities, it is necessary to look for all subsequences that represent an intrusion, which can be meant as any "improper" use of a system, an attempt to damage parts of it, to gather protected information, to follow "paths" that do not comply with security rules, etc. In this paper we propose an hypergraph-based attack model for intrusion detection. The model allows the specification of various kinds of constraints on possible attacks and provides a high degree of flexibility in representing many different security scenarios. Besides discussing the main features of the model, we study the problems of checking the consistency of attack models and detecting attack instances in sequences of logged activities.

1 Introduction

In numerous security scenarios, given a sequence of logged activities, it is necessary to look for all subsequences that represent an intrusion. The *intrusion detection* task is not generally restricted to the field of computer networks: for instance, it includes the scenarios where a surveillance system is active over a public area. An intrusion can thus be meant as any "improper" use of a system, an attempt to damage parts of it, to gather protected information, to follow "paths" that do not comply with security rules, etc.

In this paper we propose an hypergraph-based attack model for intrusion detection. Our proposed model is capable of:

- representing many different attack types/structures in a compact way;
- expressing temporal constraints on the execution of attacks;
- accommodating many different security scenarios;
- representing attack scenarios at different abstraction levels, allowing to "focus" the intrusion detection task in various ways.

An attack model is defined in such a way that the paths from start to terminal hyperedges in the model represent an attack and correspond to subsequences of a given input log—we assume that the log is a sequence of events (tuples) having a type and a timestamp. Such subsequences are the *instances* of the attack model.[1]

---

[1] Note that, although we use the word "instance", the input logs are not assumed to be generated according to our model.

M. Croitoru et al. (Eds.): GKR 2013, LNAI 8323, pp. 58–73, 2014.
© Springer International Publishing Switzerland 2014

A group of log tuples of the same type form the instance of a vertex, whereas the associations among groups form the instance of a hyperedge (called *segment*). The model allows expressing various kinds of constraints: vertices can specify type and cardinality constraints on groups of tuples and hyperedges can specify temporal constraints on the associations among groups.

It should be observed that our notion of attack model assumes that *all possible attacks* follow a path from a start to a terminal hyperedge in the model. As a matter of fact, in numerous security applications, attack models cover all possible attacks since they are defined on the basis of the specific network configuration of an organization. Activities in the model basically correspond to interdependent security vulnerabilities/alerts on specific machines – attackers *cannot* follow different paths because this would require, e.g., traversing firewalled network sections [1].

The following example intuitively introduces the main features of the proposed model.

Example 1. Consider the attack model in Fig. 1, where activities are logged security alerts and are depicted with plain circles (v_i), while hyperedges are depicted with dotted circles (h_i).

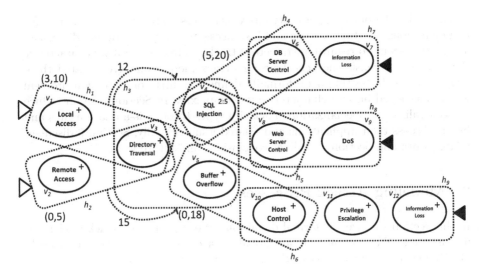

Fig. 1. Example attack model

In particular, consider hyperedges h_1, h_3, h_4, and h_7:

- h_1 is a start hyperedge (indicated with a white arrow) so an attack can begin with it. The vertex labeled **Local Access** requires the presence of a group of log tuples with one ore more tuples of type **Local Access** (cardinality constraint "+"); the same applies to the **Directory Traversal** vertex. The hyperedge itself represents an association between the two vertices, requiring that the corresponding instance (log segment) contains both the groups corresponding to **Local Access** and **Directory Traversal**, in any order. Moreover, there is a temporal constraint $(3, 10)$ that mandates, for the log segment, a temporal extension between 3 and 10 time points.

- Hyperedge h_3 requires, in any order: (i) one or more **Directory Traversal** tuples; (ii) between 2 and 5 **SQL Injection** tuples; (iii) one or more **Buffer Overflow** tuples. The temporal constraint $(0, 18)$ requires a maximum temporal extension of 18 time points for the log segment corresponding to h_3. The edge between h_1 and h_3 adds a further temporal constraint: the log tuple where the segment corresponding to h_1 begins must appear in the log at most 12 time points before the tuple where the segment for h_3 begins.
- The same applies to hyperedges h_4 and h_7. In particular, since h_7 is a terminal hyperedge (indicated with a black arrow), an attack can end with it.

Note that the absence of an annotation about the cardinality constraint of a vertex in Fig. 1 is assumed to imply that the vertex requires exactly one tuple (i.e., a "1:1" constraint). If we consider the log in Fig. 2(a), the corresponding instance is graphically described in Fig. 2(b). Further details will be provided in the next section.   □

The remainder of the paper is organized as follows. In Section 2 we give the formal definitions of attack models and instances—moreover, we briefly discuss an application scenario regarding trajectory tracking. In Section 3 we characterize the problem of consistency checking for attack models and give related theoretical results. In Section 4 we formally define the intrusion detection problem we are interested in, and characterize its complexity. In Section 5 we introduce the generalization/specialization of activity types through is-a relationships and briefly discuss their applications and related issues. Finally, Section 6 discusses related work and Section 7 outlines conclusions and future work.

2   Modeling Attack Processes

In this section we give the formal definitions of our proposed attack model and its instances. We assume that an alphabet $\mathcal{A}$ of symbols is given, univocally identifying the activity types of the underlying process of attack.

Definition 1 (Attack Model). *An* attack model *defined over the set of activity types* $\mathcal{A}$ *is a tuple* $M = \langle \mathcal{H}, \lambda, \tau, \epsilon, S, T \rangle$ *where:*

- $\mathcal{H} = (V, H)$ *is a hypergraph, where* V *is a finite set of vertices and* H *is a set of hyperedges (i.e., for each* $h \in H$, $h \subseteq V$*).*
- $\lambda : V \to \mathcal{A} \times \mathbb{N}^0 \times (\mathbb{N}^+ \cup \{\infty\})$ *is a vertex labeling function that associates with each vertex* $v \in V$ *a triple of the form* (a, l, u), *with* $l \leq u$, *which specifies the activity type of* v *along with its cardinality constraints.*[2]

---

[2] The intended meaning of the symbol '∞' is that there is no upper bound.

	ℓ_1	ℓ_2	ℓ_3	ℓ_4	ℓ_5
type	Local Access	Privilege Escalation	Buffer Overflow	Directory Traversal	SQL Injection
timestamp	0	3	5	6	10
	ℓ_6	ℓ_7	ℓ_8	ℓ_9	ℓ_{10}
type	Privilege Escalation	Buffer Overflow	SQL Injection	DB Server Control	Information Loss
timestamp	11	13	14	15	16

(a)

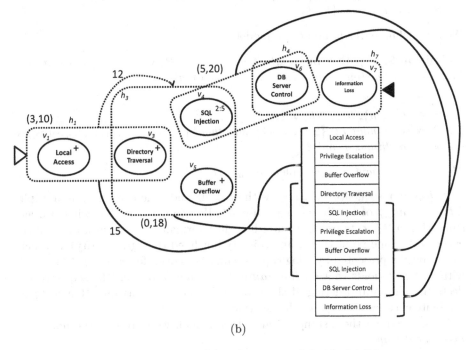

(b)

Fig. 2. Example log (a) and instance of the model (b)

- $\tau : H \rightarrow \mathbb{N}^0 \times (\mathbb{N}^+ \cup \{\infty\})$ *is a (partial) function that expresses temporal constraints (in terms of lower and upper bounds) on hyperedges—domain(τ) will denote the set of hyperedges $h \in H$ such that $\tau(h)$ is defined;*
- $\epsilon : H \times H \rightarrow \mathbb{N}^+$ *is a (partial) function that express temporal constraints (in terms of upper bounds) on ordered pairs of distinct hyperedges—domain(ϵ) will denote the set of pairs of hyperedges $h_i, h_j \in H$ such that $\epsilon(h_i, h_j)$ is defined;*
- $S, T \subseteq H$ *are nonempty sets of* start *and* terminal *hyperedges, respectively.* □

The following example shows how the model of Fig. 1 is formalized according to Definition 1.

Example 2. In the attack model $M = \langle \mathcal{H}, \lambda, \tau, \epsilon, S, T \rangle$ of Fig. 1 we have:

- $V = \{v_1, \ldots, v_{12}\}$;
- $H = \{h_1 = \{v_1, v_3\}, h_2 = \{v_2, v_3\}, h_3 = \{v_3, v_4, v_5\}, h_4 = \{v_4, v_6\}, \ldots\}$;
- $\lambda(v_1) = $ (Local Access, $1, \infty$), $\lambda(v_2) = $ (Remote Access, $1, \infty$), $\lambda(v_3) = $ (Directory Traversal, $1, \infty$), $\lambda(v_4) = $ (SQL Injection, $2, \infty$), $\lambda(v_6) = $ (DB Server Control, $1, 1$), etc.;
- $domain(\tau) = \{h_1, h_2, h_4, h_6\}$, $\tau(h_1) = (3, 10)$, $\tau(h_2) = (0, 5)$, $\tau(h_4) = (5, 20)$, $\tau(h_6) = (0, 18)$;
- $domain(\epsilon) = \{(h_1, h_3), (h_2, h_3)\}$, $\epsilon(h_1, h_3) = 12$, $\epsilon(h_2, h_3) = 15$;
- $S = \{h_1, h_2\}$, $T = \{h_7, h_8, h_9\}$. □

We now define paths in an attack model.

Definition 2. *A path π in an attack model $M = \langle \mathcal{H}, \lambda, \tau, \epsilon, S, T \rangle$ is a sequence $h_1, \ldots, h_m$ of distinct hyperedges from $\mathcal{H}$ such that*

1. *$h_1 \in S$;*
2. *$\forall i \in \{2, ..., m\}$, $h_{i-1} \cap h_i \neq \emptyset$;*
3. *there is no index $j \in \{2, ..., m-1\}$ such that $h_1, ..., h_{j-1}, h_{j+1}, ..., h_m$ satisfies both Conditions 1 and 2.*

Moreover, π is said to be complete *if $h_m \in T$.* □

A *log* is a sequence $\ell_1, ..., \ell_n$, with $n > 0$ and where each ℓ_i is a tuple $\langle att_1, \ldots, att_k \rangle$ of attributes (e.g., user-id, IP, etc.). In the following, we assume that a '*timestamp*' attribute, here just formalized as a natural number, encodes the time point (w.r.t. an arbitrary but fixed time granularity) at which the activity represented by a log tuple occurs. Moreover, for each $i, j \in \{1, ..., n\}$ with $i < j$, it holds that $\ell_i.timestamp < \ell_j.timestamp$, i.e., the sequence reflects the temporal ordering of the tuples.[3] Moreover, we assume that a '*type*' attribute encodes the type of the activity.

Before defining the instance of an attack model, we introduce the notion of *m-segmentation*.

Definition 3 (m-segmentation). *Let $L = \ell_1, ..., \ell_n$ be a log and let $m > 0$ be a natural number. A segment of L is a pair (s, t) of natural numbers such that $1 \leq s < t \leq n$. An m-segmentation of L is a sequence $(1 = s_1, t_1), ..., (s_m, t_m = n)$ of pairs of natural numbers such that:*

1. *$\forall i \in \{1, ..., m\}$, (s_i, t_i) is a segment of L;*
2. *$\forall i \in \{1, ..., m-1\}$, $s_{i+1} \leq t_i$.* □

Example 3. Consider the log L of Fig. 2(a)—for the moment, ignore the timestamps. The sequence $(1, 4), (4, 8), (5, 9), (9, 10)$ is a 4-segmentation of L that segments it into the sub-logs $L_1 = \ell_1, \ldots, \ell_4$, $L_2 = \ell_4, \ldots, \ell_8$, $L_3 = \ell_5, \ldots, \ell_9$, and $L_4 = \ell_9, \ldots, \ell_{10}$. □

---

[3] Note that we are assuming here, w.l.o.g., that there are no tuples with the same timestamp. Indeed, this can always be guaranteed by assuming a sufficiently fine time granularity.

Finally, given a log $L = \ell_1, \ldots, \ell_n$, we define the temporal distance between two tuples ℓ_i and ℓ_j in L as $d(\ell_i, \ell_j) = |\ell_j.timestamp - \ell_i.timestamp|$. We are now ready to formalize the notion of *instance* of an attack model.

An *instance* of an attack model M over a complete path in M is a log that can be segmented in such a way that, for each segment:

1. the types of the log tuples in the segment comply with the corresponding hyperedge;
2. the segment does not include unnecessary tuples as its start or end;
3. the temporal extension of the segment complies with the constraints specified by function τ (if present);
4. the start tuples of two consecutive segments comply with the constraints specified by function ϵ (if present).

The following definition formalizes this.

Definition 4 (Instance of an Attack Model). *Assume that $M = \langle \mathcal{H}, \lambda, \tau, \epsilon, S, T \rangle$ is an attack model over $\mathcal{A}$. Let $\pi = h'_1, \ldots, h'_m$ be a complete path in M, and let $L = \ell_1, \ldots, \ell_n$ be a log. Then, we say that L is an instance of M over π, denoted by $L \models_\pi M$, if there exists an m-segmentation $(s_1, t_1), \ldots, (s_m, t_m)$ of L such that $\forall i \in \{1, \ldots, m\}$,*

1. *$\forall v \in h'_i$, if $\lambda(v) = (a, l, u)$, then $l \leq |\{\ell \in \ell_s, \ldots, \ell_t | \ell.type = a\}| \leq u$;*
2. *$\exists v_s, v_t \in h'_i$ such that $\lambda(v_s) = (\ell_s.type, l_s, u_s)$ and $\lambda(v_t) = (\ell_t.type, l_t, u_t)$;*
3. *if $h'_i \in domain(\tau)$, then $l_i \leq d(\ell_{s_i}, \ell_{t_i}) \leq u_i$, with $\tau(h'_i) = (l_i, u_i)$;*
4. *$\forall i, j \in [1, m]$ s.t. $i < j$ and $(h'_i, h'_j) \in domain(\epsilon)$, it holds that $d(\ell_{s_i}, \ell_{s_j}) \leq \epsilon(h'_i, h'_j)$.* □

Example 4. Consider the attack model M of Fig. 1 and the path $\pi = h_1, h_3, h_4, h_7$ in M. The log L in Fig. 2(a) is an instance of M over π because the 4-segmentation $(1, 4), (4, 8), (5, 9), (9, 10)$ of L is such that (see Fig. 2(b)):

- the sets of activity types associated with sub-logs $L_1 = \ell_1, \ldots, \ell_4$, $L_2 = \ell_4, \ldots, \ell_8$, $L_3 = \ell_5, \ldots, \ell_9$, and $L_4 = \ell_9, \ldots, \ell_{10}$, are "minimal" supersets of the sets of activity types associated with hyperedges h_1, h_2, h_4, and h_7, respectively (Conditions 1 and 2 in Definition 4);
- temporal constraints hold (Conditions 3 and 4 in Definition 4):
  - $\tau(h_1) = (3, 10)$ and $3 \leq d(\ell_1, \ell_4) = 6 \leq 10$;
  - $\tau(h_4) = (5, 20)$ and $5 \leq d(\ell_5, \ell_9) = 5 \leq 20$;
  - $\epsilon(h_1, h_3) = 12$ and $d(\ell_1, \ell_4) = 6 \leq 12$. □

2.1 Tracking Attack Trajectories

One of the most important features of our proposed attack model is that it is general and flexible enough to support the detection of intrusions in different domains, where the activities performed by the "attacker" may not correspond to network vulnerabilities/alerts.

In fact, the only fundamental requirement of our model is the presence of timestamp and type attributes in the log. In other words, the scenarios we want to deal with must only have a temporal sequencing of activities and a finite set of activity types.

A very natural application of our model is therefore the detection of intrusions in logs that represent spatial trajectories followed by different agents/objects over time. In this case, the activity type can encode the presence of an agent/object in a predefined spatial area. As a consequence, an attack model specifies movements through areas (along with their temporal constraints) that are considered to be intrusions by the security expert. The following example shows this.

Example 5. Consider the toy attack model of Fig. 3, that tracks trajectories through airport areas—for simplicity, we do not use cardinality and temporal constraints. In this model, the attacker moves from the **Entrance** area and follows

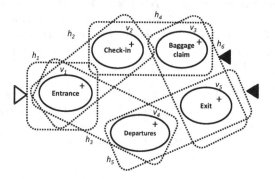

Fig. 3. Example attack model for trajectories

forbidden paths, such as **Check-in** to **Baggage claim**, **Entrance** to **Departures**, and **Departures** to **Exit**. □

3   Consistency of Attack Models

In this section we study the consistency of attack models and the complexity of checking whether a given attack model is consistent. We start by defining the consistency of a path.

Definition 5 (Consistency of a path). *Let π be a complete path in an attack model $M = \langle \mathcal{H}, \lambda, \tau, \epsilon, S, T \rangle$. We say that π is consistent w.r.t. M if there is an instance L of M over π, i.e., if there is a log L such that $L \models_\pi M$.* □

To detect the consistency of a complete path $\pi = h'_1, \ldots, h'_m$ in M, we associate a *support graph* with π, denoted by $SG(M, \pi) = \langle N, E, \omega \rangle$, that is a node- and edge-weighted directed graph where:

- the set N of nodes are the hyperedges of π;
- there is precisely an edge from h'_α to h'_β in E for each $(h'_\alpha, h'_\beta) \in domain(\epsilon)$ with $\alpha < \beta$;
- $\omega(h'_i) = \begin{cases} l_i & \text{if } h'_i \in domain(\tau) \text{ and } \tau(h'_i) = (l_i, u_i); \\ 0 & \text{if } h'_i \notin domain(\tau); \end{cases}$
- $\omega(h'_\alpha, h'_\beta) = \epsilon(h'_\alpha, h'_\beta)$.

Example 6. Consider the attack model M' obtained from the model M of Fig. 1 by adding the temporal constraint $\epsilon(h_1, h_7) = 5$. The graph $SG(M, \pi = h_1, h_3, h_4, h_7)$ is shown in Fig. 4. By definition, the graph has four nodes/hyperedges, and two edges corresponding to the two elements in the domain of ϵ. □

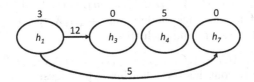

Fig. 4. Support graph associated with the path h_1, h_3, h_4, h_7

The following result gives us necessary and sufficient conditions for a path to be consistent.

Theorem 1. *Let M be an attack model and let $\pi = h'_1, \ldots, h'_m$ be a complete path in M. Then, π is consistent w.r.t. M if and only if for each edge (h'_α, h'_β) in $SG(M, \pi)$, it holds that*

$$\sum_{i=\alpha}^{\beta-1} \omega(h'_i) \le \omega(h'_\alpha, h'_\beta).$$

For instance, in Fig. 4, we have that $\epsilon(h_1, h_7) = 5 < w_{SG}(h_1) + w_{SG}(h_3) + w_{SG}(h_4) = 8$. Thus, by Theorem 1, π is not consistent w.r.t. M'.

We now define three kinds of consistency notions for attack models.

Definition 6 (Consistency of an attack model). *Assume that $M = \langle \mathcal{H}, \lambda, \tau, \epsilon, S, T \rangle$ is an attack model. We say that:*

1. *M is (S/T)-consistent if $\forall h_s \in S$, $\forall h_m \in T$ there is a path π starting with h_s and ending with h_m, respectively, that is consistent;*
2. *M is (S)-consistent if $\forall h_s \in S$, there is a complete path π starting with h_s that is consistent;*
3. *M is (T)-consistent if $\forall h_m \in T$, there is a path π terminating with h_m that is consistent.* □

Observation 1. *Let M be an attack model. If M is (S/T)-consistent, then M is both S-consistent and T-consistent.*

Finally, we characterize the complexity of checking the consistency of an attack model.

Theorem 2. *Deciding whether a given attack model is (S/T)-consistent is* **NP**-*complete, even if* $|S| = |T| = 1$.

Proof. Membership in **NP** is trivial. For the hardness, we give a reduction from the MONOTONE ONE-IN-THREE 3SAT problem, which is known to be **NP**-complete [2]. The problem is a variant of the classical satisfiability problem, where the input instance is a conjunction of clauses, with each clause consisting of exactly three variables (i.e., negation is not allowed). The goal is to determine whether there is a truth assignment to the variables so that each clause has exactly one true variable (and thus exactly two false variables).

Let $\phi = C_1 \wedge ... \wedge C_m$ be a Boolean formula taken as input and such that each clause $C_i = x_{i,1} \vee x_{i,2} \vee x_{i,3}$, $\forall i \in \{1, ..., m\}$, consists of exactly 3 variables. Based on ϕ, we define an attack model $M(\phi) = \langle \mathcal{H}, \lambda, \tau, \epsilon, S, T \rangle$ over a set $\mathcal{A}$ of activities, where $\mathcal{H} = (V, H)$ and such that:

1. $V = \{s, \phi, t, \bar{\phi}\} \cup \{C_i, x_{i,j}, \bar{C}_i, \bar{x}_{i,j} \mid i \in \{1, ..., m\} \wedge j \in \{1, 2, 3\}\}$;
2. the set H exactly contains the following hyperedges:
   - $h_s = \{s, C_1\}$;
   - $h_{i,j} = \{C_i, x_{i,j}\}$ and $h_{i,j}^{\wedge} = \{x_{i,j}, C_{i+1}\}$, $\forall i \in \{1, ..., m-1\} \wedge j \in \{1, 2, 3\}$;
   - $h_{m,j} = \{C_m, x_{m,j}\}$ and $h_{m,j}^{\wedge} = \{x_{m,j}, \phi\}$, $\forall j \in \{1, 2, 3\}$;
   - $\bar{h}_s = \{\phi, \bar{C}_1\}$;
   - $\bar{h}_{i,j} = \{\bar{C}_i, \bar{x}_{i,j}\}$ and $\bar{h}_{i,j}^{\wedge} = \{\bar{x}_{i,j}, \bar{C}_{i+1}\}$, $\forall i \in \{1, ..., m-1\} \wedge j \in \{1, 2, 3\}$;
   - $\bar{h}_{m,j} = \{\bar{C}_m, \bar{x}_{m,j}\}$ and $\bar{h}_{m,j}^{\wedge} = \{\bar{x}_{m,j}, \bar{\phi}\}$, $\forall j \in \{1, 2, 3\}$; and,
   - $h_t = \{\bar{\phi}, t\}$;
3. $\lambda(v) = (v, 1, 1)$, for each $v \in V$; in fact, $\mathcal{A} = V$;
4. $domain(\tau) = \{\bar{h}_s\}$, and $\tau(\bar{h}_s) = (2, 3)$;
5. $S = \{h_s\}$, $T = \{h_t\}$;
6. for each variable of the form $x_{i,j}$ with $i \in \{1, ..., m\}$ and $j \in \{1, 2, 3\}$, for each clause C_z with $z \in \{1, ..., m\}$ where $x_{i,j}$ occurs (possibly with $i = z$), and for each variable $x_{z,k} \neq x_{i,j}$ with $k \in \{1, 2, 3\}$ occurring in C_z and different from $x_{i,j}$, we have that: $(h_{i,j}, \bar{h}_{z,k}) \in domain(\epsilon)$ and $\epsilon(h_{i,j}, \bar{h}_{z,k}) = 1$.

As an example, the formula $\phi = (x \vee y \vee z) \wedge (x \vee y \vee w)$ is a YES instance to the MONOTONE ONE-IN-TREE 3SAT problem, as it witnessed by the truth assignment where x is the only variable evaluating true. The hypergraph associated with the attack model $M(\phi)$ is shown in Fig. 5, where for the sake of readability only arrows associated with constraints involving the variable x in ϕ are depicted.

Now, we complete the proof by claiming that: ϕ *is a YES instance to the* MONOTONE ONE-IN-THREE 3SAT *problem* $\Leftrightarrow$ *there* $M(\phi)$ *is (S/T)-consistent.*

($\Rightarrow$) Let θ be an assignment witnessing that ϕ is a YES instance. Based on θ, we build the sequence of vertices $v(\theta) = s, C_1, x_{1,j_1}, C_2, x_{2,j_2}, ..., C_m, x_{m,j_m}, \phi,$ $\bar{C}_1, \bar{x}_{1,j_1}, \bar{C}_2, \bar{x}_{2,j_2}, ..., \bar{C}_m, \bar{x}_{m,j_m}, \bar{\phi}, t$ where x_{i,j_i} is the only variable evaluating true w.r.t. θ in the clause C_i, $\forall i \in \{1, ..., m\}$. Moreover, we build the

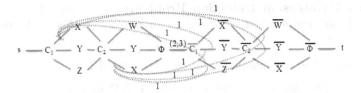

Fig. 5. The graph reduction constructed from ϕ.

sequence $\pi(\theta) = \{v_1, v_2\}, ..., \{v_i, v_{i+1}\}, ..., \{v_{4m+3}, v_{4m+4}\}$, where v_i is the i-th vertex in the sequence $v(\theta)$. Note that $\pi(\theta)$ is a path in $M(\phi)$ starting with h_s and terminating with h_t. In addition, it is consistent. Indeed, let $L_\theta = l_1, l_2, ..., l_{4m+4}$ be a log such that $l_i.type = v_i$ and $l_i.timestamp = i$, for each $i \in \{1, ..., 4m + 4\}$. Then, just notice that $L \models_{\pi(\theta)} M(\phi)$.

($\Leftarrow$) Let π be a consistent path in $M(\phi)$ starting with h_s and terminating with h_t. Note that by definition of a path, π must be of the following form: $\pi = h_s, h_{1,a_1}, h^\wedge_{1,a_1}, ..., h_{m,a_m}, h^\wedge_{m,a_m}, \bar{h}_s, \bar{h}_{1\bar{a}_1}, \bar{h}^\wedge_{1,\bar{a}_1}, ..., \bar{h}_{m,\bar{a}_m}, \bar{h}^\wedge_{m,\bar{a}_m}, h_t$, where $a_1, ..., a_m, \bar{a}_1, ..., \bar{a}_m \in \{1, 2, 3\}$. Now, recall that $\tau(\bar{h}_s) = (2, 3)$, i.e., in particular, the duration of $\bar{h}_s$ is at least 2 time units. Given the construction of ϵ, it follows that π cannot contain any pair of hyperedges in the domain of ϵ. Therefore, for each hyperedge h_{i,a_i}, with $i \in \{1, ..., m\}$, if the variable x_{i,a_i} occurs in the clause C_z, with $z \in \{1, ..., m\}$, then the hyperedge $\bar{h}_{z,\bar{a}_z}$ is actually such that $x_{z,\bar{a}_z} = x_{i,a_i}$. Let now θ be the truth assignment such that the variable x_{i,a_i} evaluates true, for each $i \in \{1, ..., m\}$ (and the remaining variables evaluate false). Note that θ is satisfying. Then, assume by contradiction that there is a clause C_z and two distinct variables $x_{z,\alpha}$ and $x_{z,\beta}$ evaluating true in θ. By construction of θ, it must be the case that $x_{z,\alpha} = x_{i',a_{i'}}$ and $x_{z,\beta} = x_{i'',a_{i''}}$, where i' and i'' are two indices in $\{1, ..., m\}$. However, by the above observations, $\bar{h}_{z,\bar{a}_z}$ is actually such that $x_{z,\bar{a}_z} = x_{i',a_{i'}} = x_{i'',a_{i''}}$. That is, $i' = i''$, which is impossible if the variables $x_{z,\alpha}$ and $x_{z,\beta}$ are distinct.

Interestingly, as the above result is obtained even when $|S| = |T| = 1$, the following result is entailed.

Corollary 1. *Deciding whether a given attack model is (S)-consistent (resp., (T)-consistent) is* **NP**-*complete.*

Present research is investigating conditions under which consistency checking becomes tractable. In fact, tractable cases may arise when there are bounds on the number and structure of hyperedges and on the size of $domain(\epsilon)$.

4   The Intrusion Detection Problem

In this section we formally characterize the intrusion detection problem we are interested in and its complexity. The problem is basically that of checking whether a log is an instance of an attack model. The following definition formalizes this.

Definition 7 (Intrusion Detection Problem). *Given an attack model M and a log L, determine whether there exists a complete path π in M such that $L \models_\pi M$.*

We now characterize the complexity of the above problem.

Theorem 3. *The intrusion detection problem is* **NP***-complete.*

Proof. Membership in **NP** is trivial: it suffices to use π and the corresponding $|\pi|$-segmentation as a polynomially-verifiable witness. We prove **NP**-hardness by polynomial-time reduction from HAMILTONIAN PATH [2]. Let $G_{in} = (V_{in}, E_{in})$ be an undirected graph. In order to decide whether G_{in} contains a Hamiltonian path, we build an attack model $M = \langle \mathcal{H}, \lambda, \tau, \epsilon, S, T \rangle$ as follows:

- $\mathcal{H} = (V, H)$;
- $V = V_{in}$;
- $H = \{\{v_1, v_2\} \mid (v_1, v_2) \in E_{in}\}$;
- $\forall v \in V, \lambda(v) = (x, 1, 1)$;
- $domain(\tau) = domain(\epsilon) = \emptyset$;
- $S = T = H$.

Then, we build a log $L = \ell_1, \ldots, \ell_n$ where $n = |V|$ and $\forall \ell_i \in L, \ell_i.type = x$ and $\ell_i.timestamp = i$. Now, in order to include all log tuples in an attack instance, all of the vertices in V must be traversed exactly once. Thus, there exists a complete path π in M such that $L \models_\pi M$ if and only if G_{in} contains a Hamiltonian path. $\qquad \square$

4.1 Scaling Intrusion Detection

Theorem 3 establishes that there are cases where detecting an intrusion is not doable in polynomial time (unless **P**=**NP**). Intuitively, the exponential blowup is mainly due to the fact that, while analyzing the log:

1. it is necessary to maintain all possible "partial" instances (prefixes of the log that are "instances" over incomplete paths) of the attack model M;
2. many different incomplete paths in M can be associated with each partial instance;
3. many different segmentations of the partial instance can be associated with each (partial instance, incomplete path) pair.

Thus, when a new log tuple is analyzed and matched against the current set of partial solutions, many new partial solutions may be generated.

   Current research is devoted to identifying conditions that make the problem tractable. One possibility is that of imposing specific limitations to the structure of functions τ and ϵ. In fact, recent works on the detection instances of temporal-automaton models in sequences of logged events [1, 3] have shown that acceptable detection times in real-world cases can be obtained by employing compact index structures and, most importantly, by limiting the number of partial solutions through a form of early filtering based on temporal constraints. In other words, they have shown that the presence of temporal constraints allows to only look at the set of current partial solutions that lie within a fixed temporal "window".

5   Type Hierarchies

In this section we show how our proposed attack model can be seamlessly extended to handle the definition of generalization/specialization hierarchies among activity types. The introduction of generalization/specialization relationships gives rise to two fundamental consequences:

- it allows to generalize/specialize an attack model when we want to track less (more) specific activity types;
- it may affect the number of instances of the model and, as a consequence, the performance of the intrusion detection task.

Fig. 6(top) shows an example hierarchy for some of the software vulnerabilities used in the attack model of Fig. 2.

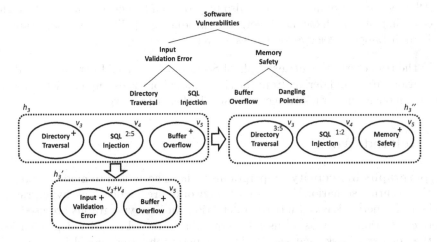

Fig. 6. Examples of abstraction based on type hierarchies

We model type hierarchies through an *is-a* relation $\Delta \subseteq \mathcal{A} \times \mathcal{A}$ and denote the transitive closure of Δ as Δ^+. With the introduction of generalization/specialization relationships, the following modifications have to be applied to Definition 1 and 4:

- In Definition 1, we add a requirement for function λ: it must be such that $\forall v, v' \in h$ with $\lambda(v) = (a, x, y)$ and $\lambda(v') = (a', x', y')$, it holds that $(a, a') \in \Delta^+$, $y' \geq x$. This ensures that the generalization/specialization relationships among the activity types in h still allow the existence of log segments that are instances of h.
- In Definition 4, in order to correctly apply generalization/specialization relationships, Conditions 1 and 2 become:
  - $\forall v \in h_i'$, if $\lambda(v) = (a, l, u)$, then $l \leq |\{\ell \in \ell_s, ..., \ell_t | (\ell.type, a) \in \Delta^+\}| \leq u$;
  - $\exists v_s, v_t \in h_i'$ such that $\lambda(v_s) = (type_s, l_s, u_s)$, $\lambda(v_t) = (type_t, l_t, u_t)$, $(\ell_s.type, type_s) \in \Delta^+$, and $(\ell_t.type, type_t) \in \Delta^+$.

The following example shows how the use of type hierarchies allows us to easily generalize/specialize an attack model and how it affects the number of instances of the model.

Example 7. Suppose we define the type hierarchy of software vulnerabilities of Fig. 6(top) and we want to modify the attack model M of Fig. 2 for an intrusion detection task where it does not matter what kind of input validation error occurs. Fig. 6(bottom) shows how we could re-design hyperedge h_3 at a higher level of abstraction: v_3 and v_4 "collapse" into a single vertex, labeled with the activity type that is the closest common ancestor of the two. The resulting edge is h_3'. The attack model M' obtained this way is a generalization of M that better fits our needs. In this case, the number of instances of M' is likely to be larger than those of M, since we now also admit instances with neither SQL Injection nor Directory Traversal activities.

A larger number of instances can also be produced by generalizing h_3 through a re-labeling of v_5 with Memory Safety, thus producing h_3''. This way, the instances will admit Dangling Pointers log tuples as well.   □

In the trajectory tracking example of Section 2.1, we could instead specialize the various areas and derive, from the single model of the example, a set of more specific models that track trajectories at a higher detail level.

6   Related Work

Hypergraphs in Security. Hypergraphs [4] have been used to model networks in the security scenario. For instance, networks are modeled as hypergraphs in [5, 6] for network-based intrusion detection. In [7], hypergraphs are used for alert correlation, whereas [8] uses hypergraphs to model security dependencies in the context of risk analysis. In [9], a hypergraph-based model is presented for describing security properties, which allow focusing on the connections between entities, by generating intrusion scenarios in situations that transcend the physical containment. In contrast to these works, we use hypergraphs to describe attacks instead of networks. A similar approach has been used in [10], where a datalog-based architecture is defined for a system that is able to reason on violations of the logging infrastructure. Instead of using an attack graph, they generalize the concept by representing the attack as a directed hypergraph, which allow to specify logical statement that describe the violation.

Workflow Modeling. Since an intrusion can be generally interpreted as a sequence of activities, many different kinds of models have been used in the past, such as, e.g., graph grammars [11] and Petri nets [12]. A number of works in the field of process modeling share with us the core idea of exploiting an hypergraph structure as generalization of a traditional (process) graph model. Unlike regular graph-structured process models, where an edge defines the sequential execution of adjacent vertices, an hyperedge is an arbitrary set of vertices which can be accomplished in any order. The applicability of hypergraphs to process modeling was studied, e.g., in [13–16], where the authors extend a metagraph structure, originally proposed in [17] to represent transformation relations between two sets of objects, into a novel model, named *metagraph-based workflow*, tailored to model workflow executions. In contrast to traditional process modeling approaches, this approach proposes a view of workflow where activities are represented by edges that relate objects (i.e., vertices of the hyperedge) that are consumed and produced during activity execution. Also inspired by hypergraph formalism, the approach in [18] uses hypergraphs to define execution semantics of *flexible process graphs* (FPG), a formal approach for modeling business processes in a flexible way. However, in a FPG, activities are represented by vertices rather than edges, and hyperedges define routing decisions by specifying executions of sets of activities that can be accomplished in an order that is actually determined when executing a process instance. Besides the specificity of the above models, the main difference with our proposed model is the focus on process modeling issues and their use for simulation/analyzing purposes. On the contrary, in our case hypergraphs are a basis for a completely new model to handle the problem of detecting intrusions.

Intrusion Detection with Attack Graphs. Our model fits well at the base of intrusion detection tasks that, to date, have often been based on direct attack graphs. In many works, such graphs are constructed by analyzing the interdependencies between vulnerabilities and security conditions that have been identified in the target network [1, 3, 19], or for correlating intrusion alerts [20–22]. One of the main advantages of using hypergraphs lies in the possibility of compactly representing a larger number of possible attack scenarios.

Finally, our idea to abstract an attack scenario using a hierarchy of activity types is related to that proposed in [23], where associating a label to each attack action is proposed, in order to report the stage where the attacker is. However, differently from our purposes, they present a quantitative threat modeling method which quantifies security threats by calculating the total severity weights of attack paths that are considered to be relevant.

7   Conclusions and Future Work

In this paper we proposed an hypergraph-based attack model for intrusion detection, showed its most important features, and studied the problems of consistency checking and attack detection. As our future work, we are currently planning to:

- Tackle the problem of scalable intrusion detection (as discussed in Section 4) and extend the intrusion detection task by assuming that (*i*) log tuples are streamed into the system in real-time and (*ii*) we want to signal the presence of a tuple with a "criticality" above the threshold as soon as it enters the system.
- Study the problems of redundancy and minimality of attack models.
- Extend the analysis to include the hierarchical levels of type (or other) attributes at subsequent stages. For instance, given a set of attack instances

computed at a high abstraction level, it could be interesting to analyze those that carried a larger "damage" at lower abstraction levels.

Acknowledgements. This work has been partially supported by the "TENACE" PRIN Project (n. 20103P34XC) and by the "Technological District on Cyber Security" PON Project, partially funded by the Italian Ministry of University and Research.

References

1. Albanese, M., Jajodia, S., Pugliese, A., Subrahmanian, V.S.: Scalable analysis of attack scenarios. In: Atluri, V., Diaz, C. (eds.) ESORICS 2011. LNCS, vol. 6879, pp. 416–433. Springer, Heidelberg (2011)
2. Garey, M.R., Johnson, D.S.: Computers and Intractability: A Guide to the Theory of NP-Completeness. W. H. Freeman & Co, New York (1979)
3. Albanese, M., Pugliese, A., Subrahmanian, V.S.: Fast activity detection: Indexing for temporal stochastic automaton-based activity models. IEEE Trans. Knowl. Data Eng. 25(2), 360–373 (2013)
4. Berge, C.: Hypergraphs: Combinatorics of Finite Sets. North-Holland (1989)
5. Vigna, G.: A topological characterization of tcp/ip security. In: Araki, K., Gnesi, S., Mandrioli, D. (eds.) FME 2003. LNCS, vol. 2805, pp. 914–939. Springer, Heidelberg (2003)
6. Vigna, G., Kemmerer, R.A.: Netstat: A network-based intrusion detection approach. In: ACSAC, pp. 25–34 (1998)
7. Morin, B., Mé, L., Debar, H., Ducassé, M.: M2D2: A formal data model for IDS alert correlation. In: Wespi, A., Vigna, G., Deri, L. (eds.) RAID 2002. LNCS, vol. 2516, pp. 115–127. Springer, Heidelberg (2002)
8. Baiardi, F., Suin, S., Telmon, C., Pioli, M.: Assessing the risk of an information infrastructure through security dependencies. In: López, J. (ed.) CRITIS 2006. LNCS, vol. 4347, pp. 42–54. Springer, Heidelberg (2006)
9. Pieters, W.: Ankh: Information threat analysis with actor-network hypergraphs. CTIT technical report series, Enschede, Centre for Telematics and Information Technology, University of Twente (2010)
10. Johnson, C.R., Montanari, M., Campbell, R.H.: Automatic management of logging infrastructure. In: National Centers of Academic Excellence - Workshop on Insider Threat, St Louis, MO, USA (2010)
11. Korff, M., Ribeiro, L.: Formal relationship between graph grammars and petri nets. In: Cuny, J., Engels, G., Ehrig, H., Rozenberg, G. (eds.) Graph Grammars 1994. LNCS, vol. 1073, pp. 288–303. Springer, Heidelberg (1996)
12. Alimonti, P., Feuerstein, E.: Petri nets, hypergraphs and conflicts (preliminary version). In: Mayr, E.W. (ed.) WG 1992. LNCS, vol. 657, pp. 293–309. Springer, Heidelberg (1993)
13. Basu, A., Blanning, R.W.: Metagraphs in workflow support systems. Decision Support Systems 25(3), 199–208 (1999)
14. Basu, A., Blanning, R.W.: A formal approach to workflow analysis. Information Systems Research 11(1), 17–36 (2000)
15. Basu, A., Blanning, R.W.: Workflow analysis using attributed metagraphs. In: HICSS (2001)

16. Basu, A., Blanning, R.W.: Metagraphs and Their Applications. Integrated Series in Information Systems. Springer, Dordrecht (2007)
17. Basu, A., Blanning, R.W.: Metagraphs: a tool for modeling decision support systems. Manage. Sci. 40(12), 1579–1600 (1994)
18. Polyvyanyy, A., Weske, M.: Hypergraph-based modeling of ad-hoc business processes. In: Ardagna, D., Mecella, M., Yang, J. (eds.) BPM 2008 Workshops. LNBIP, vol. 17, pp. 278–289. Springer, Heidelberg (2009)
19. Ammann, P., Wijesekera, D., Kaushik, S.: Scalable, graph-based network vulnerability analysis. In: ACM Conference on Computer and Communications Security, pp. 217–224 (2002)
20. Noel, S., Robertson, E., Jajodia, S.: Correlating intrusion events and building attack scenarios through attack graph distances. In: ACSAC, pp. 350–359 (2004)
21. Wang, L., Liu, A., Jajodia, S.: An efficient and unified approach to correlating, hypothesizing, and predicting intrusion alerts. In: De Capitani di Vimercati, S., Syverson, P.F., Gollmann, D. (eds.) ESORICS 2005. LNCS, vol. 3679, pp. 247–266. Springer, Heidelberg (2005)
22. Wang, L., Noel, S., Jajodia, S.: Minimum-cost network hardening using attack graphs. Computer Communications 29(18), 3812–3824 (2006)
23. Chen, Y., Boehm, B.W., Sheppard, L.: Value driven security threat modeling based on attack path analysis. In: HICSS, p. 280 (2007)

Structural Consistency: A New Filtering Approach for Constraint Networks*

Philippe Jégou and Cyril Terrioux

LSIS - UMR CNRS 7296
Aix-Marseille Université
Avenue Escadrille Normandie-Niemen
13397 Marseille Cedex 20, France
{philippe.jegou,cyril.terrioux}@lsis.org

Abstract. In this paper, we introduce a new partial consistency for constraint networks which is called *Structural Consistency* of level w and is denoted w-*SC* consistency. This consistency is based on a new approach. While conventional consistencies generally rely on local properties extended to the entire network, this new partial consistency considers global consistency on subproblems. These subproblems are defined by partial constraint graphs whose tree-width is bounded by a constant w. We introduce a filtering algorithm which achieves w-SC consistency. We also analyze w-SC filtering w.r.t. other classical local consistencies to prove that this consistency is generally incomparable although this consistency can be regarded as a special case of inverse consistency. Finally, we present experimental results to assess the usefulness of this approach. We show that w-SC is a significantly more powerful level of filtering and more effective w.r.t. the runtime than SAC and that w-SC is a complementary approach to AC or SAC. So we can offer a combination of filterings, whose power is greater than w-SC or SAC.

1 Introduction

It is well known that the CSP formalism (Constraint Satisfaction Problems [1]) is important in the field of AI to express and then to efficiently solve a large class of problems. A CSP, also called constraint network, consists of a set of variables X, each of which must be assigned a value in its associated finite domain from D, and a solution must satisfy a finite set C of constraints. Classical approaches to find solutions are based on backtracking algorithms whose time complexity cost is $O(e.a.d^n)$ where n is the number of variables, e is the number of constraints, a denotes a bound on constraint arity, and d is the maximum size of domains, assuming that the cost of a constraint check is $O(a)$. To efficiently solve CSP, algorithms use classically filtering techniques, before search as preprocessing or during search. The quality of this basic tool is generally crucial for the efficiency of the search.

---

* This work was supported by the French National Research Agency under grant TUPLES (ANR-2010-BLAN-0210).

M. Croitoru et al. (Eds.): GKR 2013, LNAI 8323, pp. 74–91, 2014.

The effect of filtering techniques often consists in removing values from domains. These values can be safely deleted because they cannot appear in solutions (they are not consistent). So, filterings are based on the notion of consistency. Because removing all inconsistent values is generally unrealistic from a practical viewpoint (it is an NP-hard problem), filterings are based on local consistencies which are relaxed consistency properties. So, a value can satisfy a local consistency, even if it does not appear in any solution. Nevertheless other values can contradict partial consistency and then be removed without modifying the satisfiability of a CSP. Partial consistencies [2] are mainly defined by local consistencies which are extended to the whole constraint network. For example, to satisfy arc-consistency (AC), the most popular local consistency, a value needs to possess at least one compatible value (called a support) in the domain of its neighbouring variables. Otherwise, this value is removed from its domain (filtered) and this deletion can produce other deletions in the domains of its neighbouring variables. By using a mechanism called constraint propagation, the first deletion can finally be extended to the whole network. So, partial consistencies are generally local properties which must be verified on the whole network. From a practical viewpoint, the interest of a partial consistency is related to its filtering power and to the cost for enforcing it (time and space complexities).

In this paper, we introduce a new kind of inverse consistency which must satisfy two criteria, particularly, on hard instances:

- practical efficiency,
- filtering power.

For that, this new consistency will be:

- parametrized, to control the complexity of the filtering,
- adjusted to the structure of the constraint network, by exploiting its substructures,
- adjusted to the tightness of the constraints in selecting the tightest constraints.

This new consistency is called w-SC because it is a parametrized Consistency (w is the parameter) based on Structural properties of the network. It is defined on a relaxation of the considered CSP (a subproblem) which is a partial constraint network whose tree-width is bounded by a constant w [3]. We choose this kind of subproblems because they can be solved in polynomial time (in w). Furthermore, thanks to recent progress on decomposition methods, they can be managed really efficiently from a practical viewpoint [4]. Moreover, while classical partial consistencies consider constraints independently from their tightness, here we can select subsets of constraints focusing particularly on tight constraints when selecting the considered subgraph. More precisely, given a sub-network corresponding to a partial graph of bounded tree-width, we will say that a value satisfies w-SC consistency if it appears at least in one solution of the considered subproblem. w-SC consistency can be considered as a new case of inverse consistency but it is formally different from inverse consistencies already known [5]. So, this new consistency allows us to define a new kind of filtering which is finally different from filterings generally used in CSPs. Notably, we will show that w-SC is incomparable with existing efficient partial consistencies such as AC or SAC [6]. Particularly, experiments show that the behavior of w-SC consistency is different from the behavior of these consistencies, being more efficient for hard instances, while being less efficient on easy (unconstrained instances). Moreover, we have observed that the values deleted by w-SC are generally not deleted by AC or SAC, and the reverse is true. Hence, w-SC can be considered as a local consistency which is a complementary consistency of already known consistencies. So we propose a combination of filterings based on SAC and w-SC whose power is significantly greater than w-SC or SAC. We also show that w-SC is a significantly more powerful level of filtering and more effective w.r.t. the runtime than SAC.

This paper is organized as follows. The next section recalls classical notions on partial consistencies and their related filterings. Then, in section 3, we introduce w-SC consistency and its associated filtering. Section 4 presents a theoretical analysis of relations between w-SC consistency and other consistencies, while section 5 provides an empirical analysis of this filtering. Finally, the last section is devoted to a discussion about future works.

2   Preliminaries

A *finite constraint satisfaction problem* or *finite constraint network* (X, D, C) is defined as a set of variables $X = \{x_1, \ldots x_n\}$, a set of domains $D = \{D(x_1), \ldots D(x_n)\}$ (the domain $D(x_i)$ contains all the possible values for the variable x_i), and a set C of constraints among variables. A constraint $c_i \in C$ is defined by its scope, denoted $S_C(c_i)$ and by an associated relation $R_C(c_i)$. The scope is an ordered subset of variables, that is $S_C(c_i) = (x_{i_1}, x_{i_2}, \ldots x_{i_{a_i}})$ where a_i is called the *arity* of the constraint c_i. The relation $R_C(c_i) \subseteq D(x_{i_1}) \times D(x_{i_2}) \ldots \times D(x_{i_{a_i}})$ defines the allowed combinations of values for the variables in $S_C(c_i)$. Here, we denote by S_C the set of scopes of the constraints, that is $S_C = \{Sc_1, Sc_2 \ldots Sc_e\}$ where $e = |C|$ is the number of constraints. A solution of (X, D, C) is an assignment of each variable which satisfies all the constraints. Without loss of generality, we assume that each variable is involved in at least one constraint. If every constraint of a CSP is binary (i.e. involves exactly two variables), then the structure of this binary network (called a binary CSP) can be represented by the graph (X, S_C) called the *constraint graph*.

In this paper, we assume that the relations are not empty. Moreover, without loss of generality, we will assume that the constraint network is connected and normalized (two different constraints do not involve the same variables) and for sake of simplicity, we will consider here only binary networks. So, for a constraint c_k such $S_C(c_k) = (x_i, x_j)$, c_k will be denoted c_{ij}. Generally, CSPs are solved using backtracking algorithms which can be really efficient if they efficiently exploit filterings before or during search to avoid redundant search. These filterings are formally based on the notion of local consistency.

The most popular and oldest local consistency is called *arc-consistency* (AC). Given a CSP $P = (X, D, C)$, a value $v_i \in D(x_i)$ is arc-consistent w.r.t. $c_{ij} \in C$

iff there exists a valid value $v_j \in D(x_j)$ s.t. $(v_i, v_j) \in R_C(c_{ij})$. Then, $v_j \in D(x_j)$ is a *support* of v_i for the constraint c_{ij}. A domain $D(x_i)$ is arc-consistent w.r.t. c_{ij} iff it is not empty and $\forall v_i \in D(x_i)$, the value v_i is arc-consistent w.r.t. c_{ij}, and the CSP $P = (X, D, C)$ is arc-consistent iff $\forall D(x_i) \in D$, the domain $D(x_i)$ is arc-consistent w.r.t. all $c_{ij} \in C$. A filtering of domains based on AC consists in removing the values which do not satisfy arc-consistency. When a value v_i is removed, a mechanism called *constraint propagation* can be run to remove values which were supported only by v_i (no other value of $D(x_i)$ is compatible with them), and this process can be extended to other values. For binary networks, many efficient algorithms have been proposed (e.g. AC-2001 [7]) to enforce arc-consistency. Their practical efficiency makes it possible to use them during search. From the theoretical viewpoint, the best algorithms (in terms of worst-case time complexity) have a time complexity in $O(e.d^2)$. Nevertheless, the filtering power of AC can be really limited because of the local definition of the consistency. So, more powerful consistencies performing more powerful filterings have been defined. [8] has introduced *k-consistency* which considers subsets of k variables. For k-consistency, a new constraint (its arity is $k - 1$) is added to the network when a consistent assignment on $k - 1$ variables cannot be extended to a kth variable. If the network is i-consistent, for $2 \leq i \leq k$, the CSP is said *strong k-consistent*. The greater the value of k is, the more powerful the filtering is. Unfortunately, the time and space complexity is $O(n^k.d^k)$ [9] and then this kind of filtering has important drawbacks. Because of the time and space complexity, these filterings are generally unusable even for small values of k ($k = 3$ is frequently unrealistic for practical cases). Moreover, the filtering will add new constraints in the network, their arity being $k - 1$ and then the necessary space can be prohibitive, even for small values of k. To avoid these problems, mainly the second one related to added constraints, other local consistencies have been introduced. For example, [5] have proposed the *k-inverse consistency*. The associated filtering removes values which cannot be extended to any $k - 1$ additional variables to form a consistent assignment. This filtering is more powerful than AC and it avoids the problem related to space complexity but for time complexity, the problem remains the same since it is $O(n^k.d^k)$. So, they have suggested to limit k-inverse consistency to small values of k. In [5], another kind of inverse consistency has been introduced which is defined in the same spirit and which is called *NIC* for *neighborhood-inverse consistency*. Here, the filtering of a domain is induced by the compatibility of values of the associated variable w.r.t. the subproblem defined by its neighborhood in the network. So, the complexity is related to the maximum degree Δ of a variable in the constraint network, and then the time complexity of the proposed algorithm is $O(\Delta^2.(n+e.d).d^{\Delta+1})$. Another way to define local consistencies is based on the subproblem induced by an assignment $x_i = v_i$. For example, a CSP is *Singleton arc-consistency* (denoted *SAC*) [6] if for all domains and then all their values $v_i \in D(x_i)$, the subproblem induced by the assignment $x_i = v_i$ has arc-consistent sub-domains. The time complexity is $O(e.n.d^3)$ [10].

To conclude this overview, we must recall that from a practical viewpoint, the local consistencies generally considered as the most usable in practice are AC or SAC which seem to obtain the best compromise between the time cost (and its practical efficiency) and its filtering power.

3   Structural Consistency

3.1   *w*-SC Consistency

Structural consistency is based on the notion of partial graph whose *tree-width* is bounded by a constant w. The tree-width is based on the notion of *tree-decomposition* which has been formally introduced in [3]. It has been exploited in the field of CSP to define tractable classes [11] and to propose efficient methods for solving constraint networks which possess good topological properties [12,4].

Our objective here is then different from these works since we will use it to define local consistencies. For that, we introduce the notion of *w-PST* which corresponds to *partial spanning tree-decomposition* of tree-width w. Before, we recall the classical notion of tree-decomposition:

Definition 1. *A tree-decomposition of a graph* $G = (X, E)$ *is a pair* (N, T) *where* $T = (I, F)$ *is a tree with nodes* I *and edges* F *and* $N = \{N_i : i \in I\}$ *is a family of subsets of* X, *s.t. each subset (called cluster)* N_i *is a node of* T *and verifies:*

(i) $\cup_{i \in I} N_i = X$,
(ii) for each edge $\{x, y\} \in E$, *there exists* $i \in I$ *with* $\{x, y\} \subseteq N_i$, *and*
(iii) for all $i, j, k \in I$, *if* k *is in a path from* i *to* j *in* T, *then* $N_i \cap N_j \subseteq N_k$.
The width w *of a tree-decomposition* (N, T) *is equal to* $max_{i \in I} |N_i| - 1$. *The tree-width* w^* *of* G *is the minimal width over all the tree-decompositions of* G.

Now, we define partial graphs with particular tree-width.

Definition 2. *Given a graph* $G = (V, E)$, *a partial graph of* G *is a subgraph* $G' = (V, E')$ *where* $E' \subseteq E$. *A partial spanning tree-decomposition of tree-width* w *for* G, *denoted* w-PST, *is a partial graph of* G *whose tree-width is* w.

The graph given in figure 1(a) is a 3-PST because its tree-width is 3 (an optimal tree-decomposition is given in this figure). In figure 1(b), we have the partial graph induced by the deletion of edges $\{2, 4\}$ and $\{9, 10\}$ in the graph given in figure 1(a). It is a 2-PST because its tree-width is 2 as indicated by the optimal tree-decomposition given in this figure.

Now, we introduce the notion of subproblem of a given CSP induced by the assignment of a variable.

Definition 3. *Given a CSP* $P = (X, D, C)$, *a variable* $x_i \in X$ *and a value* $v_i \in D(x_i)$, *the subproblem of* P *induced by the assignment* $x_i = v_i$ *is* $P|_{x_i = v_i} = (X, D', C')$ *where* $D'(x_i) = \{v_i\}$ *and for all* $j \neq i, D'(x_j) = D(x_j)$ *and* $C' = C$ *except for the relations associated to the constraints including* x_i *which are restricted to the tuples where the value* v_i *appears.*

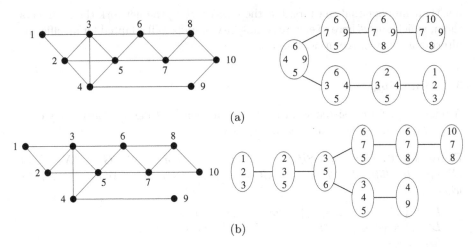

Fig. 1. (a) A graph of tree-width 3 and the corresponding tree-decomposition. (b) A 2-PST of the graph given in (a) and an optimal tree-decomposition.

We introduce now the notion of relaxed problem of a given CSP which is defined by a subset of constraints.

Definition 4. *Given a CSP $P = (X, D, C)$ and $W \subseteq S_C$, the relaxed problem of P induced by W is $P(W) = (X, D, C')$ where $W = S_{C'}$.*

For w-SC, the relaxed problem is defined by a subset of constraints forming a partial spanning tree-decomposition of tree-width w.

Definition 5. *Given a CSP $P = (X, D, C)$ and a w-PST $G = (X, W)$ of (X, S_C):*

- *The value $v_i \in D(x_i)$ is w-SC-consistent w.r.t. G iff $P(W)|_{x_i = v_i}$ has a solution.*
- *The domain $D(x_i)$ is w-SC-consistent w.r.t. G iff $\forall v_i \in D(x_i)$, the value v_i is w-SC-consistent w.r.t. G.*
- *The CSP $P = (X, D, C)$ is w-SC-consistent w.r.t. G iff $\forall D(x_i) \in D$, $D(x_i)$ is w-SC-consistent w.r.t. G.*

Note that for a given CSP and a given value w, there are different possible consistencies, since one consistency is defined with respect to one particular w-PST. Moreover, this definition of consistency is related to values of domains but it can be easily generalized to partial assignments of subsets of variables. In order to simplify our presentation, we limit here w-SC-consistency to values.

As we can see, the w-SC-consistency is an inverse consistency. It may therefore be related to works as those of [5]. One can also consider the Relational (i, m)-consistency [13] where i would be equal to 1 since we filter domains. However, compared with [13], the parameter m plays here a significantly different role since it is related to the number of constraints. For the w-SC-consistency, it is not their number that matters, but the tree-width of the network that connects them, which allows here to ensure complexity bounds well adapted for an efficient filtering as we will see in next section.

3.2 Filtering

As classically for consistencies, the filtering associated to w-SC consistency consists in deleting values which do not satisfy it.

Definition 6. *Given a CSP $P = (X, D, C)$ and a w-PST $G = (X, W)$ of (X, S_C), the filtered CSP using w-SC-consistency is w-SC$(P, W) = (X, D', C')$ where:*

- $D' = \{D'(x_1), \ldots D'(x_n)\}$ *where* $\forall D'(x_i) \in D'$,
  $D'(x_i) = \{v_i \in D(x_i): v_i$ *is w-SC-consistent w.r.t. $G\}$.*
- $S_{C'} = S_C$.
- $\forall c'_{ij} \in C', R_{C'}(c'_{ij}) = R_C(c_{ij}) \cap D'(x_i) \times D'(x_j)$.

Note that given a CSP and a w-PST (X, W) of (X, S_C), w-SC(P, W) is unique. Moreover, to ensure that w-SC consistency defines a valid filtering, we must also ensure that no filtered value can appear in solutions of the given CSP. It is necessarily the case since removed values cannot appear in solutions of a relaxed CSP. Now, we present the algorithm called *Comp-w-SC* which achieves w-SC filtering. Contrary to classical filtering algorithms as those enforcing AC, this consistency does not need a propagation step after deletions. Indeed, while classical algorithms remove values and propagate these deletions, here once a value v_i is validated finding a solution, it will not be deleted after, and thus v_i is definitively validated. It is because v_i appears in a solution of the relaxed CSP, and because the other values that appear in this solution (which can be considered as supports for the value v_i) are also validated by the same reason. Thus, since these values are also validated, it will not be necessary to check their w-SC consistency.

In *Comp-w-SC*, the function *Solution$(P(W)|_{x_i=v_i}, Sol)$* is called to check consistency. If v_i appears in a solution $Sol = (v_1, v_2, \ldots v_i, \ldots v_n)$ of $P(W)$, the function returns *true* and *Sol* is the other result of this call. Otherwise, it returns *false*. In *Comp-w-SC*, D' corresponds to the set of domains containing values which have already been validated and then memorized during the filtering. So, if a value v_i already appears in $D'(x_i)$ it is because this value already appears in a solution and then it will not be necessary to check it after for its w-SC consistency. Note that at the end of *Comp-w-SC*, for all variable x_i, we have $D(x_i) = D'(x_i)$.

Property 1. *The time complexity of Comp-w-SC is $O(n^2.w.d^{w+2})$ while its space complexity is $O(n.w.d^w)$.*

Proof: The function *Solution$(P(W)|_{x_i=v_i})$* is called at most $n.d$ times and the cost of one call to this function is bounded by $n.w.d^{w+1}$. Indeed, it can be

---

Algorithm 1: Comp-w-SC(**In:** (X, W): Graph; **InOut:** $P = (X, D, C)$: CSP)

---

1   **for** $x_i \in X$ **do**
2      $D'(x_i) \leftarrow \emptyset$;

3   **for** $x_i \in X$ **do**
4   **for** $v_i \in D(x_i)$ **do**
5   **if** $v_i \notin D'(x_i)$ **then**
6   **if** $Solution(P(W)|_{x_i=v_i}, Sol)$ **then**
7   **for** $v_j \in Sol$ **do**
8                 $D'(x_j) \leftarrow D'(x_j) \cup \{v_j\}$
9   **else**
10               $D(x_i) \leftarrow D(x_i) - \{v_i\}$

---

implemented using algorithms based on tree-decomposition of CSPs such as TC [12] or BTD [4].

Moreover, we know that the space complexity of decomposition methods as BTD is related to the size of the separators between cliques [4]. Here, the maximum size of separators in the w-PST is w and their number is at most $n - 1$. So, the space complexity is bounded by $O(n.w.d^w)$.   □

Note that the algorithm Comp-w-SC is really similar to the algorithm 1 proposed in [14]. Nevertheless, in this paper, the motivations of the authors to realize a filtering is related to the number of solutions which is too important and then they want to obtain a minimal network w.r.t. the domains of values. Moreover, they suppose that their problem is easy to solve and then, they do not indicate this cost to evaluate their algorithm. Here, we propose to bound the time complexity of this algorithm by a polynomial related to the value of w.

4   Relations with Other Consistencies

To evaluate the power of the w-SC filtering, we provide here an analysis in the same spirit as in [6] which presents the comparison between numerous partial consistencies. Here, we consider AC, SAC, strong-PC, and more generally strong-k-consistency and k-inverse consistency. The comparison is based on formal relations between consistencies that we recall now. We say that a consistency CO_1 is *stronger* than a consistency CO_2 (denoted $CO_2 \le CO_1$) if in any CSP instance P in which CO_1 holds, CO_2 holds too. So, any algorithm achieving CO_1 deletes at least the values removed by CO_2. We say that a consistency CO_1 is *strictly stronger* than a consistency CO_2 (denoted $CO_2 < CO_1$) if $CO_2 \le CO_1$ and there is at least one CSP instance P in which CO_2 holds and CO_1 does not. Note that these relations are transitive. Finally, we say that CO_1 and CO_2 are *incomparable* if neither relation between them hold.

Note that for a given CSP and a given value w, the number of possible filterings is potentially related to the number of possible w-PSTs. Nevertheless, we can easily find instances of CSPs such that next properties hold.

Theorem 1. *1-SC < AC and for $w > 1$, w-SC and AC are incomparable.*

Proof: It is clear that connected 1-PST are exactly trees. So, since in a tree, a value appears in a solution iff it verifies AC, 1-SC cannot filter more values than the arc-consistency, which considers the whole problem, does. Thus, we have 1-SC $\leq$ AC. Moreover, since other values of a network can be deleted by AC, exploiting constraints that does not appear in the considered 1-PST, we have also 1-SC < AC.

Now, if we consider w-SC and AC for $w > 1$, they are incomparable. Indeed, it is sufficient to see that if a value in the domain of a variable is removed because it has no support for a constraint which does not appear in a w-PST, then this value can be conserved by w-SC filtering. Conversely, a value can be removed by w-SC filtering but not by AC. □

Theorem 2. *1-SC < SAC and for $w > 1$, w-SC and SAC are incomparable.*

Proof: Since 1-SC < AC and since AC < SAC, by transitivity of <, we have 1-SC < SAC. Now, since SAC considers all the constraints which appear in the network, necessarily, the filtering can delete values that will not be deleted by 2-SC. Conversely, if we consider the example (c) given in page 216 of [6] which is a 2-PST satisfying SAC, we can easily see that 2-SC will remove the value deleted by strong-PC while this value is not deleted by SAC. Thus, 2-SC and SAC are incomparable. □

Before comparing stronger consistencies, we define relations between different levels of w-SC consistency. For relations between w-SC with different values of w, we assume that we consider CSPs which possess partial spanning tree-decompositions with different widths such that w-PST $\subseteq$ $(w + 1)$-PST (edges of the considered w-PST satisfy this condition). Otherwise we have no guarantee about the comparison between w-SC and $(w + 1)$-SC.

Theorem 3. *If w-PST $\subseteq (w + 1)$-PST, then w-SC < $(w + 1)$-SC.*

Proof: If w-PST $\subseteq (w+1)$-PST, the values removed by w-SC consistency considering the w-PST are necessary removed considering the $(w + 1)$-PST. Moreover, since there are constraints in the $(w + 1)$-PST that do not appear in the w-PST, additional values will be removed by $(w + 1)$-SC consistency in considering the $(w + 1)$-PST. Thus, we have w-SC < $(w + 1)$-SC. □

More generally, applying the same principle as for the theorem 1, we have the next property that can be considered as its generalization:

Theorem 4. *k-SC < strong-$(k+1)$-consistency and for $k > 2$, k-SC and strong-k-consistency are incomparable.*

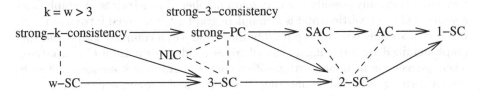

Fig. 2. Relations between consistencies

Proof: Firstly, we show that k-SC $<$ strong-$(k+1)$-consistency. It is easy to see that the width [15] of a w-PST is exactly w. Consequently, applying the results proposed in [15] which links the width of a constraint network to the level of strong-consistency that it verifies, necessarily, every value that appears in a domain satisfying strong-$(k+1)$-consistency belongs to a solution of this k-PST. Thus, it will not be deleted by k-SC consistency. Moreover, since other values of the network can be deleted by strong-$(k+1)$-consistency, exploiting constraints that do not appear in the considered w-PST (with $w = k$), it is easy to see that strong-$(k+1)$-consistency can delete more values than k-SC consistency filtering does, and consequently, we have k-SC $<$ strong-$(k+1)$-consistency.

To show that for $k > 2$, k-SC and strong-k-consistency are incomparable, it is then sufficient to show that some values can be deleted by k-SC, while they are not filtered by strong-k-consistency. Consider the CSP corresponding to the k-coloring problem on the complete graph on $k + 1$ vertices. This network is a k-PST but it does not admit a solution. Thus, achieving of k-SC consistency on this network will delete all values. Now, if we consider strong-k-consistency, no value will be deleted. Consequently, for $k > 2$, k-SC and strong-k-consistency are incomparable. □

So, we have the next trivial corollary:

Corollary 1. *2-SC $<$ strong-PC.*

Finally, note that we can replace strong-k-consistency by k-inverse-consistency [5]. It is also easy to establish that NIC and k-SC (for $k > 2$) are incomparable.

In figure 2, relations between consistencies are summarized. Arrows represents the relation $>$ while dotted lines indicates that two consistencies are incomparable.

5 Experiments

5.1 Experimental Protocol

The quality of the filtering performed by w-SC mainly depends on the considered w-PST. The computation of the w-PST is achieved thanks to a heuristic method related to the notion of k-tree [16]. We exploit k-trees here because they are the graphs which, for a given tree-width, have the maximum number of edges, and thus can potentially contain the maximum number of constraints. A graph G is a k-tree if G has k vertices and is a complete graph (it is a trivial k-tree) or there is a vertex x of degree k whose neighborhood induces a complete graph and the graph obtained by removing x and all edges incident to it from G is a k-tree. Then, k-trees can be built starting with a complete graph of k vertices, and each time a vertex x is added, connecting it to vertices of a k-clique in the previous graph. In our case, we choose at each step the vertex x which has the smallest value $\prod_{c_{ij}} t_{ij}$ where c_{ij} is a constraint spanned by the clique formed by x and the vertices of the considered k-clique in the previous graph and t_{ij} is the ratio of the number of forbidden tuples over the number of possibles tuples. Likewise the initial clique is chosen in a similar way. By so doing, we aim to compute tight subproblems and so obtain a more powerful filtering. We also implement another method which computes first a k-tree as previously described and then tries heuristically to add as many constraints (among the constraints which do not belong to the k-tree yet) as possible such that the tree-width remains bounded by w. The remaining constraints are processed in the decreasing tightness order. In the following results, w-SC1 (respectively w-SC2) denotes the application of our algorithm with a w-PST computed thanks to the first method (resp. thanks to the second one). In both cases, we exploit a k-tree with $k = w$ and the subproblem related to the considered w-PST is solved thanks to BTD [4].

Regarding AC and SAC, we have implemented AC2001 [7] and a naive version of SAC. All the algorithms are written in C. Note that we have also compared our results with a clever implementation of SAC (namely SAC3 [17] provided in the Java solver Abscon). Regarding the runtime, our implementation performs sometimes worse than SAC3 but the results of the comparison with w-SC remain the same. Of course, both versions of SAC have the same filtering power. So, in order to make easier the implementation of the combination of SAC and w-SC (see subsection 5.3), we only consider our implementation in the provided results.

These algorithms are compared on random instances produced by the random generator written by D. Frost, C. Bessière, R. Dechter and J.-C. Régin. Note that we do not use here structured random instances because, even if w-SC takes benefit from the underlying structure of the CSP, it aims to be run on general CSP instances as most filtering algorithms. This generator takes 4 parameters n, d, e and T. It builds a CSP of class (n, d, e, t) with n variables which have domains of size d and e binary constraints ($0 \leq e \leq \frac{n(n-1)}{2}$) in which t tuples are forbidden ($0 \leq t \leq d^2$). The presented results are the averages of the results obtained on 50 instances (with a connected constraint graph) per class (except for figures 4 and 5 where we consider only 30 instances per class). The experimentations are performed on a linux-based PC with an Intel Pentium IV 3.2 GHz and 1 GB of memory.

5.2   AC/SAC vs w-SC

We have tested many classes of instances by varying the number of variables, the size of domains (up to 40 values), the constraint graph density and the tightness.

Table 1. Runtime (in ms), number of instances detected as inconsistent and mean number of removed values. All the considered instances have no solution.

Classes (n,d,e,t)	AC time	#inc	#rv	SAC time	#inc	#rv	6-SC1 time	#inc	#rv	6-SC2 time	#inc	#rv
(a) (100,20,495,275)	1.8	0	9.24	198	50	104.68	70	20	486.82	441	48	79.20
(b) (100,20,990,220)	2.4	0	0.22	11987	11	92.04	105	21	566.08	240	48	79.32
(c) (100,20,1485,190)	3.4	0	0	4207	0	0.40	286	31	494.40	187	49	38.30
(d) (100,40,495,1230)	4.6	0	0.92	3239	50	345.06	270	21	621.94	5709	48	106.76
(e) (100,40,990,1030)	5.8	0	0	13229	0	0.08	515	30	809.34	4954	48	176.42
(f) (100,40,1485,899)	8.2	0	0	11166	0	0	1622	32	902.14	1854	48	181.18
(g) (200,10,1990,49)	2.6	0	20.96	88	50	38.28	128	22	350.62	72	48	56.36
(h) (200,10,3980,35)	5.8	0	0.92	10503	49	261.74	248	0	34.86	637	0	249.56
(i) (200,10,5970,30)	7.8	0	0.24	11335	0	11	423	0	54.62	708	2	241.34
(j) (200,20,995,290)	4.6	0	57.58	224	50	65.52	190	26	670.32	7464	49	78.42
(k) (200,20,1990,245)	6	0	3.36	3716	50	256.62	192	32	592.96	1109	50	20
(l) (200,20,3980,195)	12.4	0	0.04	34871	0	1.82	573	25	808.46	592	49	70.48
(m) (200,20,5970,165)	17	0	0	23307	0	0.04	2242	10	1179.88	1600	43	280.3

Figures 3-5 and table 1 present some representative results obtained on various classes. Note that we do not provide the results of w-SC2 in figures 3-5 because the results are very close to ones of w-SC1 (w-SC2 only detects the inconsistency of a few additional instances while it spends a slightly greater time).

Before comparing our algorithm with AC or SAC, we raise the question of the choice of a good value for w. If we consider the number of instances which are detected as inconsistent, we can note that this number for w-SC increases as the value of w increases. Such a result is foreseeable since for larger values of w, w-SC takes into account more constraints and is able to perform a more powerful filtering. The same result generally holds for the runtime which increases with w. According to our observations, the value 6 for w seems to correspond to the best trade-off between the runtime and the power of the filtering. On the one hand, by exploiting 6-PSTs, 6-SC1 (or 6-SC2) takes into account enough constraints in order to enforce an efficient filtering. On the other hand, with larger values of w, the number of removed values and so the number of detected inconsistent instances are not significantly improved while the runtime and the space requirement may increase significantly w.r.t. those for $w = 6$.

Then, if we compare w-SC1 and w-SC2, we can observe in table 1 that generally w-SC2 detects more instances as inconsistent than w-SC1. Again, such a result was foreseeable, since w-SC2 takes into account more constraints than w-SC1. For the same reason, w-SC2 often spends more time for achieving SC.

Now, if we compare the w-SC filtering with AC or SAC w.r.t the number of instances which are detected as inconsistent, we note that 3-SC detects more inconsistencies than AC while SAC may perform better or worse than w-SC depending on the value of w. Nevertheless, we can observe that 6-SC often detects more instances as inconsistent than SAC. On the 650 instances considered in table 1, we observe that, SAC performs often better than 6-SC1 (310 instances are detected as inconsistent by SAC against 270 by 6-SC1) but worse than 6-SC2 (310 against 530).

Regarding the runtime, according to figure 3, AC is generally faster than w-SC, except when the instances are obviously inconsistent. In such a case, w-SC often succeeds in detecting the inconsistency by removing all the values of the first considered variable while AC needs many deletions. Compared with SAC, w-SC spends more time than SAC only on instances which are not tight enough (e.g. for $t < 50$ in figure 3) and so obviously consistent, because many calls to BTD are required. In contrast, when the tightness is closer to the consistent/inconsistent threshold or above, w-SC performs faster than SAC. Indeed, w-SC removes less values than SAC in order to detect the inconsistency. The main reason is that, by construction, it checks (and possibly removes) each value of a variable before considering a new variable while in AC or SAC, the values of a given variable are generally deleted at different (and non-consecutive) moments (due to the propagation mechanism). Finally, we have observed that the behavior of w-SC with respect to AC or SAC is improved when the density of the constraint graph increases. So, w-SC succeeds in outperforming significantly SAC for time efficiency and AC and SAC for detection of inconsistencies in the consistent/inconsistent threshold area.

5.3   Complementarity and Combinations of AC/SAC and w-SC

If we compare the values which are deleted by the different algorithms on consistent instances, we can note that a value deleted by w-SC is not necessarily removed by AC or SAC and conversely. Such a report highlights the difference which exists between our new filtering and classical ones like AC or SAC and leads us to study the complementarity of w-SC with AC or SAC. With this aim in view, we combine here w-SC with AC or SAC. More precisely, from two filtering algorithms X and Y, we derive a new filtering algorithm denoted X+Y which consists in applying X and then, if the instance is not detected as inconsistent yet, Y. Table 2 provides the results obtained by AC+6-SC, SAC+6-SC, 6-SC+AC and 6-SC+SAC.

We can note that AC+6-SC slightly improves the detection of inconsistent instances w.r.t 6-SC (e.g. 275 instances are detected as inconsistent by AC+6-SC1 against 270 for 6-SC1) while 6-SC+AC performs really better with 450 instances found inconsistent by 6-SC1+AC (respectively 567 by 6-SC2+AC against 530 for 6-SC2). Both algorithms have better results than AC which do not succeed in detecting the inconsistency of any instance. Likewise, SAC+6-SC improves both the behaviour of SAC and 6-SC (457 and 587 instances respectively for SAC+6-SC1 and SAC+6-SC2 against 310 for SAC). For these three algorithms, the runtime does not exceed the cumulative runtime of AC and 6-SC or SAC and 6-SC. The more interesting results are provided by 6-SC+SAC. On the one hand, 6-SC1+SAC and 6-SC2+SAC detect respectively 598 and 640 instances as inconsistent (i.e. 92% and 98% of the considered instances against 48% for SAC, 42% for 6-SC1 and 82% for 6-SC2). On the other hand, its runtime is often significantly better than the cumulative runtime of SAC and 6-SC or than the runtime of SAC. These results clearly show that the filtering performed by AC/SAC and SC are not only significantly different but also complementary.

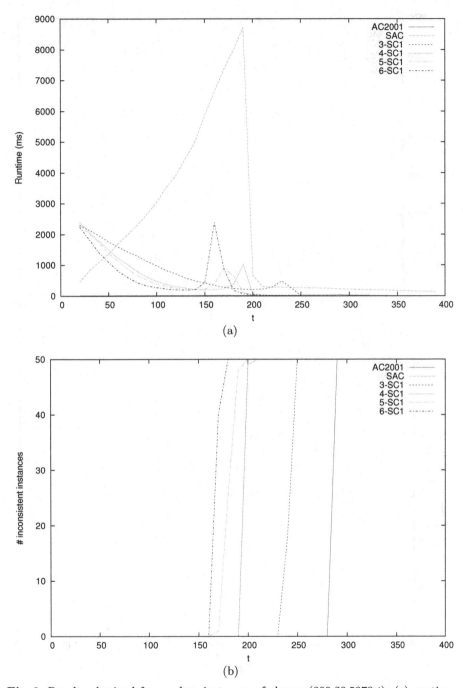

Fig. 3. Results obtained for random instances of classes (200,20,5970,*t*): (a) runtime in ms and (b) number of detected inconsistent instances

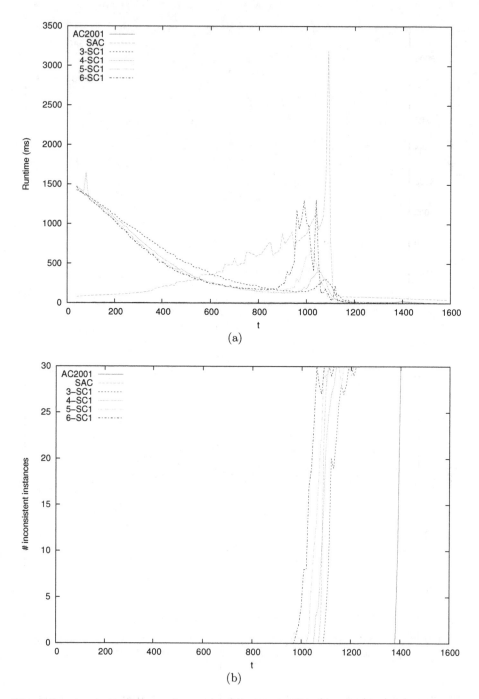

Fig. 4. Results obtained for random instances of classes (100,40,990,t) with 30 instances per class: (a) runtime in ms and (b) number of detected inconsistent instances

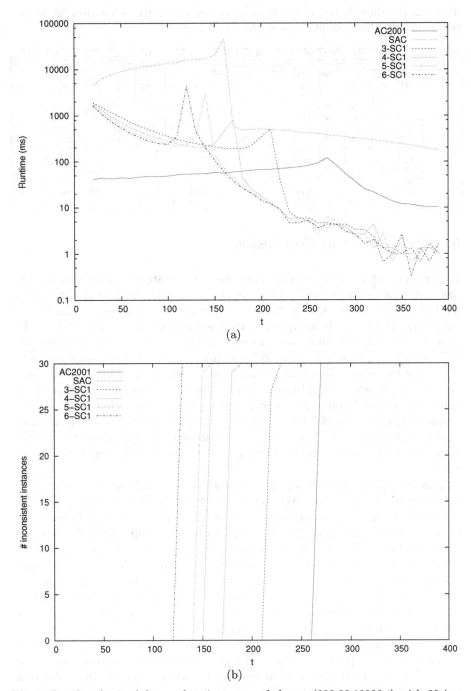

Fig. 5. Results obtained for random instances of classes (200,20,19900,t) with 30 instances per class: (a) runtime in ms (with a logarithmic scale) and (b) number of detected inconsistent instances

Table 2. Runtime (in ms) and number of instances detected as inconsistent for the combined algorithms (for SC, we consider 6-SC1 and 6-SC2)

Classes	AC+SC1		SAC+SC1		SC1+AC		SC1+SAC		AC+SC2		SAC+SC2		SC2+AC		SC2+SAC	
(n,d,e,t)	time	#inc	time	#inc	time	#inc	time	#inc	time	#inc	time	#inc	time	#inc	time	#inc
(a)	70	20	197	50	71	41	86	50	381	48	198	50	386	50	430	50
(b)	104	21	12020	28	100	40	1387	49	238	48	12035	48	240	50	223	50
(c)	272	31	4487	32	288	47	525	48	203	49	4385	49	202	50	185	50
(d)	248	21	3249	50	268	30	608	50	5724	48	3240	50	4950	49	5754	50
(e)	526	30	13835	30	579	41	1729	48	4982	48	18407	48	5546	50	5590	50
(f)	1629	32	12822	32	1631	44	2466	48	1858	48	13015	48	1826	50	2024	50
(g)	125	25	89	50	133	45	136	50	67	50	91	50	73	50	73	50
(h)	261	0	10715	49	260	0	6946	49	654	0	10536	49	638	10	2701	50
(i)	424	0	15939	0	424	1	15744	18	711	2	16060	3	728	12	5906	42
(j)	175	28	225	50	199	45	191	50	5215	50	223	50	7933	50	7965	50
(k)	186	32	3774	50	181	45	352	50	1086	50	3614	50	1114	50	1133	50
(l)	578	25	35291	26	539	42	3130	49	627	49	35352	49	593	50	549	50
(m)	2253	10	25526	10	2235	29	15216	39	1696	43	24923	43	1592	46	5056	48

6   Discussion and Conclusion

We have proposed a new parameterized partial consistency which allows to exploit the underlying properties of a constraint network, both at the structural level but also at the level of the tightness of constraints. This new partial inverse consistency is different in its approach to those proposed previously. This can be seen in the theoretical comparisons we have provided, but also in the experiments since its filtering capabilities are different from those of conventional methods: this is not the same values that are removed, and time efficiency is not located in the same areas as other filterings. It is therefore a potentially complementary approach to existing ones. In experimental results, we show that w-SC offers a significantly more powerful level of filtering and more effective w.r.t the runtime than SAC. We also show that w-SC is a complementary approach to AC or SAC. So we can offer a combination of filterings, whose power is greater than w-SC or SAC.

This leads naturally to study hybrid consistencies combining w-SC with other consistencies such as AC or SAC, to provide filtering both more robust and more powerful. Among the potential extensions of this work, w-SC could be extended to n-ary constraints, which does not seem to be technically difficult. We could also extend w-SC in extending filterings to partial assignments, for example by offering the concept of k-w-SC consistency which would produce new constraints whose arity is k. Moreover, even if 6-SC is faster and performs a more powerful filtering than SAC, it would also be necessary to better identify the good values of w. A natural track would be to assess the density of the w-PST for a given value of w. Another important way could be to use w-SC during the search. We should study both conventional methods based on backtracking, but also to improve the decomposition methods. Finally, a comprehensive study should be conducted based on the general framework proposed in [18] by studying different kinds of subproblems, not only related to partial spanning tree-decompositions. Nevertheless, this study could be of a limited interest. Indeed, this framework allows to define a generic algorithm which is based on a classic propagation architecture. Given a deleted value, this value will be propagated to delete other values. Conversely, achieving w-SC is not based on deletion propagation but runs using validations of value. This difference makes the use of the framework of [18] of a limited interest here for filtering.

References

1. Rossi, F., van Beek, P., Walsh, T.: Handbook of Constraint Programming. Elsevier (2006)
2. Bessiere, C.: Constraint Propagation. In: Rossi, F., van Beek, P., Walsh, T. (eds.) Handbook of Constraint Programming, ch. 3, pp. 29–83. Elsevier (2006)
3. Robertson, N., Seymour, P.D.: Graph minors II: Algorithmic aspects of treewidth. Algorithms 7, 309–322 (1986)
4. Jégou, P., Terrioux, C.: Hybrid backtracking bounded by tree-decomposition of constraint networks. Artificial Intelligence 146, 43–75 (2003)
5. Freuder, E., Elfe, C.D.: Neighborhood inverse consistency preprocessing. In: Proceedings of AAAI, pp. 202–208 (1996)
6. Debruyne, R., Bessière, C.: Domain Filtering Consistencies. JAIR 14, 205–230 (2001)
7. Bessière, C., Régin, J.C., Yap, R.H.C., Zhang, Y.: An optimal coarse-grained arc consistency algorithm. Artificial Intelligence 165(2), 165–185 (2005)
8. Freuder, E.: Synthesizing constraint expressions. CACM 21(11), 958–966 (1978)
9. Cooper, M.C.: An optimal k-consistency algorithm. Artificial Intelligence 41(1), 89–95 (1989)
10. Bessière, C., Debruyne, R.: Optimal and suboptimal singleton arc consistency algorithms. In: Proceedings of IJCAI, pp. 54–59 (2005)
11. Gottlob, G., Leone, N., Scarcello, F.: A Comparison of Structural CSP Decomposition Methods. Artificial Intelligence 124, 243–282 (2000)
12. Dechter, R.: Constraint processing. Morgan Kaufmann Publishers (2003)
13. Dechter, R., van Beek, P.: Local and Global Relational Consistency. Journal of Theoretical Computer Science (1996)
14. Bayer, K.M., Michalowski, M., Choueiry, B.Y., Knoblock, C.A.: Reformulating CSPs for Scalability with Application to Geospatial Reasoning. In: Bessière, C. (ed.) CP 2007. LNCS, vol. 4741, pp. 164–179. Springer, Heidelberg (2007)
15. Freuder, E.: A Sufficient Condition for Backtrack-Free Search. JACM 29(1), 24–32 (1982)
16. Beineke, Pippert: Properties and characterizations of k-trees. Mathematika 18, 141–151 (1971)
17. Bessière, C., Cardon, S., Debruyne, R., Lecoutre, C.: Efficient Algorithms for Singleton Arc Consistency. Constraints 16(1), 25–53 (2011)
18. Verfaillie, G., Martinez, D., Bessière, C.: A Generic Customizable Framework for Inverse Local Consistency. In: Proceedings of AAAI, pp. 169–174 (1999)

Inductive Triple Graphs: A Purely Functional Approach to Represent RDF

Jose Emilio Labra Gayo[1], Johan Jeuring[2], and Jose María Álvarez Rodríguez[3]

[1] University of Oviedo, Spain
labra@uniovi.es
[2] Utrecht University, Open University of the Netherlands, The Netherlands
j.t.jeuring@uu.nl
[3] South East European Research Center, Greece
jmalvarez@seerc.org

Abstract. RDF is one of the cornerstones of the Semantic Web. It can be considered as a knowledge representation common language based on a graph model. In the functional programming community, inductive graphs have been proposed as a purely functional representation of graphs, which makes reasoning and concurrent programming simpler. In this paper, we propose a simplified representation of inductive graphs, called Inductive Triple Graphs, which can be used to represent RDF in a purely functional way. We show how to encode blank nodes using existential variables, and we describe two implementations of our approach in Haskell and Scala.

1 Introduction

RDF appears at the basis of the semantic web technologies stack as the common language for knowledge representation and exchange. It is based on a simple graph model where nodes are predominantly resources, identified by URIs, and edges are properties identified by URIs. Although this apparently simple model has some intricacies, such as the use of blank nodes, RDF has been employed in numerous domains and has been part of the successful linked open data movement.

The main strengths of RDF are the use of global URIs to represent nodes and properties and the composable nature of RDF graphs, which makes it possible to automatically integrate RDF datasets generated by different agents.

Most of the current implementations of RDF libraries are based on an imperative model, where a graph is represented as an adjacency list with pointers, or an incidence matrix. An algorithm traversing a graph usually maintains a state in which visited nodes are collected.

Purely functional programming offers several advantages over imperative programming [13]. It is easier to reuse and compose functional programs, to test properties of a program or prove that a program is correct, to transform a program, or to construct a program that can be executed on multi-core architectures.

M. Croitoru et al. (Eds.): GKR 2013, LNAI 8323, pp. 92–110, 2014.
© Springer International Publishing Switzerland 2014

In this paper, we present a purely functional representation of RDF Graphs. We introduce popular combinators such as fold and map for RDF graphs. Our approach is based on Martin Erwig's inductive functional graphs [10], which we have adapted to the intricacies of the RDF model. The main contributions of this paper are:

- a simplified representation of inductive graphs
- a purely functional representation of RDF graphs
- a description of Haskell and Scala implementations of an RDF library

This paper is structured as follows: Section 2 describes purely functional approaches to graphs. In particular, we present inductive graphs as introduced by Martin Erwig, and we propose a new approach called triple graphs, which is better suited to implement RDF graphs. Section 3 presents the RDF model. Section 4 describes how we can represent the RDF model in a functional programming setting. Section 5 describes two implementations of our approach: one in Haskell and another in Scala. Section 6 describes related work and Section 7 concludes and describes future work.

2 Inductive Graphs

2.1 General Inductive Graphs

In this section we review common graph concepts and the inductive definition of graphs proposed by Martin Erwig [10].

A directed graph is a pair $\mathcal{G} = (\mathcal{V}, \mathcal{E})$ where $\mathcal{V}$ is a set of vertices and $\mathcal{E} \subseteq \mathcal{V} \times \mathcal{V}$ is a set of edges. A labeled directed graph is a directed graph in which vertices and edges are labeled. A vertex is a pair (v, l), where v is a node index and l is a label; an edge is a triple (v_1, v_2, l) where v_1 and v_2 are the source and target vertices and l is the label.

Example 21. *Figure 1 depicts the labeled directed graph with* $\mathcal{V} = \{(1, a), (2, b), (3, c)\}$, *and* $\mathcal{E} = \{(1, 2, p), (2, 1, q), (2, 3, r), (3, 1, s)\}$.

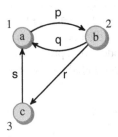

Fig. 1. Simple labeled directed graph

In software, a graph is often represented using an imperative data structure describing how nodes are linked by means of edges. Such a data structure may be an adjacency list with pointers, or an incidence matrix. When a graph changes, the corresponding data structure is destructively updated. A graph algorithm that visits nodes one after the other uses an additional data structure to register what part of the graph has been visited, or adapts the graph representation to include additional fields to mark nodes and edges in the graph itself.

Implementing graph algorithms in a functional programming language is challenging as one has to either pass an additional parameter to all the functions with that data structure or use monads to encapsulate the imperative style. This style complicates correctness proofs and program transformations.

Martin Erwig [9] introduces a functional representation of graphs where a graph is defined by induction. He describes two implementations that enable persistent graphs [8], and an implementation in Haskell [10], which we summarize in this section. He defines a graph as either 1) an empty graph or 2) an extension of a graph with a node v together with its label and a list of v's succesors and predecessors that are already in the graph.

The type of the values used in an extension of a graph is given by the type Context.

```
1  -- Context of a node in the graph
2  type Context a b =
3      (Adj b, Node, a, Adj b)
4
5  -- Adjacent labelled nodes
6  type Adj b = [(Node,b)]
7
8  -- Labelled nodes
9  type LNode a = (a,Node)
10
11  -- Index of nodes
12  type Node = Int
13
14  -- Labelled edges
15  type LEdge b = (Node,Node,b)
```

A context of a node is a value (pred, node, lab, succ), where pred is the list of predecessors, node is the index of the node, lab is the label of the node and succ is the list of successors. Labelled nodes are represented by a pair consisting of a label and a node, and labelled edges are represented by a source and a target node, together with a label.

Example 22. *The context of node b in Figure 1 is:*

```
1  ([(1,'p')],2,'b',[(1,'q'),(3,'r')])
```

Although the graph type is implemented as an abstract type for efficiency reasons, it is convenient to think of the graph type as an algebraic type with two constructors Empty and :&.

```
1  data Graph a b = Empty
2     | Context a b :& Graph a b
```

Example 23. *The graph from Figure 1 can be encoded as:*

```
1  ([(2,'q'),(3,'s')],1,'a',[(2,'p')]) :&
2  ([],2,'b',[(3,'r')]) :&
3  ([],3,'c',[]) :&
4  Empty
```

Note that there may be different inductive representations for the same graph.

Example 24. *Here is another representation of the graph in Figure 1:*

```
1  ([(2,'r')],3,'c',[(1,'s')]) :&
2  ([(1,'p')],2,'b',[(1,'q')]) :&
3  ([],1,'a',[]) :&
4  Empty
```

The inductive graph approach has been implemented in Haskell in the FGL library[1]. FGL defines a type class Graph to represent the interface of graphs and some common operations. The essential operations are:

```
1  class Graph gr where
2    empty:: gr a b
3    isEmpty:: gr a b -> Bool
4    match:: Node -> gr a b ->
5        (Context a b, gr a b)
6    mkGraph::[LNode a] -> [LEdge b]
7        -> gr a b
8    labNodes :: gr a b -> [LNode a]
```

Fig. 2. Inductive graph representation using M. Erwig approach

A problem with this interface is that it exposes the management of node/edge indexes to the user of the library. It is for example possible to construct graphs with edges between non-existing nodes.

Example 25. *The following code compiles but produces a runtime error because there is no node with index 42:*

---

[1] http://web.engr.oregonstate.edu/~erwig/fgl/haskell

```
1  gErr :: Gr Char Char
2  gErr = mkGraph
3    [('a',1)]
4    [(1,42,'p')]
```

2.2 Inductive Triple Graphs

We propose a simplified representation of inductive graphs based on three assumptions:

- each node and each edge have a label
- labels are unique
- the label of an edge can also be the label of a node

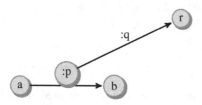

Fig. 3. A triple graph with an edge acting also as a node

These three assumptions are motivated by the nature of RDF Graphs, which we will explain in the next section. As we will see in Section 2.3, our approach is general enough to represent any graph.

One advantage of this representation is that a user does not have to be aware of node indexes. Also, there is no need for two different types for nodes/edges simplifying the development of an algebra of graphs.

A graph of elements of type a is described by a set of triples where each triple has the type (a,a,a). We will call these kind of graphs TGraph (triple based graphs).

We assume triple graphs are defined by the following datatype. Practical implementations may use a different representation.

```
1  data TGraph a = Empty
2               | TContext a :& TGraph a
```

where TContext a is defined as:

```
1  type TContext a =
2    (a, [(a,a)], [(a,a)], [(a,a)])
```

A TContext of a node is a value (node,pred,succ,rels) where node is the node itself, pred is the list of predecessors, succ is the list of successors and rels is the list of pairs of nodes related by this node when it is an edge.

Example 26. *The graph from Figure 1 can be defined as:*

```
1  ('a',[('c','s'),('b','q')],
2       [('p','b')],
3       []) :&
4  ('b',[],[('r','c')],[]) :&
5  ('c',[],[],[]) :&
6  ('p',[],[],[]) :&
7  ('q',[],[],[]) :&
8  ('r',[],[],[]) :&
9  ('s',[],[],[]) :&
10 Empty
```

With this representation it is easy to model graphs in which edges are also nodes.

Example 27. *The graph from Figure 3 can be defined by:*

```
1  ('a',[],[('p','b')],[]) :&
2  ('b',[],[],[]) :&
3  ('p',[],[('q','r')],[]) :&
4  ('q',[],[],[]) :&
5  ('r',[],[],[]) :&
6  Empty
```

As in Erwig's approach, it is possible to have different representations for the same graph.

Example 28. *The previous graph can also be defined as follows, where we reverse the order of the nodes:*

```
1  ('r',[],[('p','q')],[]) :&
2  ('q',[],[],[]) :&
3  ('p',[],[],[('a','b')]) :&
4  ('b',[],[],[]) :&
5  ('a',[],[],[]) :&
6  Empty
```

In Haskell, we implement TGraph as a type class with the following essential operations:

Using this simplified interface, it is impossible to create graphs with edges between non-existing nodes.

2.3 Representing Graphs at Triple Graphs

We can represent general inductive graphs [10] using inductive triple graphs. The main difference between general inductive graphs and inductive triple graphs is that in general inductive graphs, labels of nodes and edges have an index (an Int), which does

```
1  class TGraph gr where
2  -- empty graph
3  empty :: gr a
4
5  -- decompose a graph
6  match :: a -> gr a -> (TContext a,gr a)
7
8  -- make graph from triples
9  mkGraph :: [(a,a,a)] -> gr a
10
11  -- nodes of a graph
12  nodes :: gr a -> [a]
13
14  -- extend a graph (similar to :&)
15  extend :: TContext a -> gr a -> gr a
```

Fig. 4. TGraph representation

not need to be different. We represent a general inductive graph using a record with a triple graph that stores either the index of the node or the index of the edge, and two maps, one from indexes to node labels and another from indexes to edge labels.

```
1  data GValue a b = Node a | Edge b
2
3  data Graph a b = Graph {
4  graph :: TGraph (GValue Int Int),
5  nodes :: Map Int a
6  edges :: Map Int b
```

Example 29. *The graph from example 24 can be represented as:*

```
1  Graph {
2  graph =
3   (Node 1,[(Node 3,Edge 4),
4           (Node 2,Edge 2)],
5           [(Edge 1,Node 2)],
6           [] :&
7   (Node 2,[],
8           [(Edge 3,Node 3)],
9           []) :&
10  (Node 3,[],[],[]) :&
11  (Edge 1,[],[],[]) :&
12  (Edge 2,[],[],[]) :&
13  (Edge 3,[],[],[]) :&
14  (Edge 4,[],[],[]) :&
15  Empty,
```

```
16    nodes = Map.fromList
17       [(1,'a'),(2,'b'),(3,'c')],
18    edges = Map.fromList
19       [(1,'p'),(2,'q'),(3,'r'),(4,'s')]
20  }
```

The conversion between both representations is straightforward and is available at `https://github.com/labra/haws`.

Conversely, we can also represent inductive triple graphs using general inductive graphs. As we describe in Section 5, our Haskell implementation is defined in terms of Martin Erwig's FGL library.

2.4 Algebra of Graphs

Two basic operators on datatypes are the *fold* and the *map* [17] . The fold is the basic recursive operator on datatypes: any recursive function on a datatype can be expressed as a fold. Using the representation introduced above, we can define `foldGraph`:

```
1  foldTGraph :: TGraph gr =>
2    b -> (TContext a -> b -> b) ->
3                       gr a -> b
4  foldTGraph e f g = case nodes g of
5    [] -> e
6    (n:_) -> let (ctx,g') = match n g
7             in f ctx (foldTGraph e f g')
```

The map operator applies an argument function to all values in a value of a datatype, preserving the structure. It is the basic functorial operation on a datatype. On `TGraph`'s, it takes a function that maps a-values in the context to b-values, and preserves the structure of the argument graph. We define `mapGraph` in terms of `foldGraph`.

```
1  mapTGraph :: TGraph gr =>
2    (TContext a -> TContext b) ->
3                       gr a -> gr b
4  mapTGraph f =
5    foldTGraph empty
6      (\ctx g -> extend (mapCtx f ctx) g)
7    where
8      mapCtx f (n,pred,succ,rels) =
9       (f n,
10       mapPairs f pred,
11       mapPairs f succ,
12       mapPairs f rels)
13      mapPairs f = map
14       (\(x,y) -> (f x, f y))
```

An interesting property of `mapTGraph` is that it maintains the graph structure whenever the function f is injective. Otherwise, the graph structure can be completely modified.

Example 210. *Applying the function `mapTGraph (\_ -> 0)` to a graph returns a graph with a single node.*

Using `mapGraph`, we define some common operations over graphs.

Example 211. *The following function reverses the edges in a graph.*

```
1  rev :: (TGraph gr) => gr a -> gr a
2  rev = mapTGraph swapCtx
3   where
4    swapCtx (n,pred,succ,rels) =
5     (n,succ,pred,map swap rels)
```

We have defined other graph functions implementing depth-first search, topological sorting, strongly connected components, etc. [2]

3   The RDF Model

The RDF Model was accepted as a recommendation in 2004 [1]. The 2004 recommendation is being updated to RDF 1.1, and the current version [5] is the one we use for the main graph model in this paper. Resources in RDF are globally denoted IRIs (internationalized resource identifiers [7]).[3] Notice that the IRIs in the RDF Model are global identifiers for nodes (subjects or objects of triples) and for edges (predicates). Therefore, an IRI can be both a node and an edge. Qualified names are employed to shorten IRIs. For example, if we replace `http://example.org` by the prefix `ex:`, `ex:a` refers `http://example.org/a`. Throughout the paper we will employ Turtle notation [6]. Turtle supports defining triples by declaring prefix aliases for IRIs and introducing some simplifications.

Example 31. *The following Turtle code represents the graph in Figure 1.*

```
1  @prefix : <http://example.org/>
2
3  :a :p :b .
4  :b :q :a .
5  :b :r :c .
6  :c :s :a .
```

---

[2] The definitions can be found on `https://github.com/labra/haws`.
[3] The 2004 RDF recommendation employs URIs, but the current working draft uses IRIs.

An RDF triple is a three-tuple $\langle s, p, o \rangle \in (\mathcal{I} \cup \mathcal{B}) \times \mathcal{I} \times (\mathcal{I} \cup \mathcal{B} \cup \mathcal{L})$, where $\mathcal{I}$ is a set of IRIs, $\mathcal{B}$ a set of blank nodes, and $\mathcal{L}$ a set of literals. The components s, p, o are called, the subject, the predicate, and the object of the triple, respectively. An RDF graph $\mathcal{G}$ is a set of RDF triples.

Example 32. *The following Turtle code represents the graph in Figure 3.*

```
7   :a  :p  :b  .
8   :p  :q  :r  .
```

Blank nodes in RDF are used to describe elements whose IRI is not known or does not exist. The Turtle syntax for blank nodes is _:id where id represents a local identifier for the blank node.

Example 33. *The following set of triples can be depicted by the graph in Figure 5.*

```
9    :a  :p  _:b1  .
10   :a  :p  _:b2  .
11   _:b1  :q  :b  .
12   _:b2  :r  :b  .
```

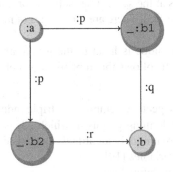

Fig. 5. Example with two blank nodes

Blank node identifiers are local to an RDF document and can be described by means of existential variables [16]. Intuitively, a triple $\langle b_1, p, o \rangle$ where $b_1 \in \mathcal{B}$ can be read as $\exists b_1 \langle b_1, p, o \rangle$. This predicate holds if there exists a resource s such that $\langle s, p, o \rangle$ holds.

When interpreting an RDF document with blank nodes, arbitrary resources can be used to replace the blank nodes, replacing the same blank node by the same resource.

Currently, the RDF model only allows blank nodes to appear as subjects or objects, and not as predicates. This restriction may be removed in future versions of RDF so we do not impose it in our graph representation model. Literals are used to denote values such as strings, numbers, dates, etc. There are two types of literals: datatype literals and language literals. A datatype literal is a pair (val, t) where $val \in \mathcal{L}$ is a lexical form representing its value and $t \in \mathcal{T}$ is a datatype URI. In Turtle, datatype literals are represented as `val^^t`. A language literal is a pair $(s, lang)$ where $s \in \mathcal{L}$ is a string value and $lang$ is a string that identifies the language of the literal.

In the RDF data model, literals are constants. Two literals are equal if their lexical form, datatype and language are equal. The different lexical forms of literals can be considered unique values. Although the current RDF graph model restricts literals to appear only as objects, we do not impose that restriction in our model. For simplicity, we only use lexical forms of literals in the rest of the paper.

4   Functional Representation of RDF Graphs

An RDF document in the RDF model is a labeled directed graph where the nodes are resources. A resource can be modeled as an algebraic datatype:

```
1  data Resource = IRI String
2               | Literal String
3               | BNode BNodeId
4
5  type BNodeId = Int
```

The RDF graph model has three special aspects that we need to take into account:

- edges can also be nodes at the same time (subjects or objects)
- nodes are uniquely identified. There are three types of nodes: resource nodes, blank nodes and literals
- the identifier of a blank node is local to the graph, and has no meaning outside the scope of the graph. It follows that a blank node behaves as an existential variable [16]

To address the first two aspects we employ the triple inductive graphs introduced in Section 2.2, which support defining graphs in which edges can also appear as nodes, and both nodes and edges are uniquely identified. The existential nature of blank nodes can be modeled by logical variables [19].

The type of RDF graphs is defined as:

```
1  data RDFGraph = Ground (TGraph Resource)
2               | Exists (BNodeId -> RDFGraph)
```

Example 41. *The graph from Figure 5 is defined as:*

```
1  Exists (\b1 ->
2  Exists (\b2 ->
3  Ground (
4    ('a',[],[('p',b1),('p',b2)],[]) :&
5    ('b',[(b1,'q'),(b2,'r')],[],[]) :&
6    (b1, [], [], []) :&
7    (b2, [], [], []) :&
8    (p, [], [], []) :&
```

```
 9      (q, [], [], []) :&
10      (r, [], [], []) :&
11      Empty)))
```

This RDFGraph encoding makes it easy to construct a number of common functions on RDF graphs. For example, two RDFGraph's can easily be merged by means of function composition and folds over triple graphs.

```
1  mergeRDF :: RDFGraph -> RDFGraph -> RDFGraph
2  mergeRDF g (Exists f) = Exists (\x -> mergeRDF g (f x))
3  mergeRDF g (Ground g') = foldTGraph g compRDF g'
4   where
5    compRDF ctx (Exists f) =
6         Exists (\x -> compRDF ctx (f x))
7    compRDF ctx (Ground g) =
8         Ground (comp ctx g)
```

We define the map function over RDFGraphs by:

```
1  mapRDFGraph::(Resource -> Resource) ->
2                   RDFGraph -> RDFGraph
3  mapRDFGraph h (Basic g) =
4       Basic (gmapTGraph (mapCtx h) g)
5  mapRDFGraph h (Exists f) =
6       Exists (\x -> mapRDFGraph h (f x))
```

Finally, to define foldRDFGraph, we need a seed generator that assigns different values to blank nodes. In the following definition, we use integer numbers starting from 0.

```
1  foldRDFGraph ::
2    a -> (Context Resource -> a -> a) ->
3         RDFGraph -> a
4  foldRDFGraph e h =
5    foldRDFGraph' e h 0
6   where
7    foldRDFGraph' e h seed (Ground g) =
8      foldTGraph e h g
9    foldRDFGraph' e h seed (Exists f) =
10     foldRDFGraph' e h (seed+1) (f seed)
```

5 Implementation

We have developed two implementations of inductive triple graphs in Haskell[4]: one using higher-order functions and another based on the FGL library. We have also developed a Scala implementation[5] using the *Graph for Scala* library.

5.1 Implementation in Haskell

Our first implementation uses a functional representation of graphs. A graph is defined by a set of nodes and a function from nodes to contexts.

```
data FunTGraph a =
 FunTGraph (a -> Maybe (Context a, FunTGraph a))
          (Set a)
```

This implementation offers a theoretical insight but is not intended for practical proposes.

The second Haskell implementation is based on the FGL library. In this implementation, a `TGraph a` is represented by a `Graph a` and a map from nodes to the edges that they relate.

```
data FGLTGraph a = FGLTGraph {
  graph :: Graph a a,
  nodeMap :: Map a (ValueGraph a)
}

data ValueGraph a = Value {
 grNode :: Node,
 edges :: Set (a,a)
}
```

`nodeMap` keeps track of the index of each node in the graph and the set of (subject,object) nodes that the node relates if it acts as a predicate. Any inductive triple graph can be converted to an inductive graph using Martin Erwig's approach.

5.2 Implementation in Scala

In Scala, we define a `Graph` trait with the following interface:

```
trait TGraph[A] {
  def empty : TGraph[A]

  def mkTGraph
      (triples : Set((A,A,A))): TGraph[A]

```

---

[4] Haskell implementations are available at https://github.com/labra/haws.
[5] Scala implementation is available at https://github.com/labra/wesin.

```
7   def nodes : Set[A]

8

9   def decomp
10        (node : A): (Context[A],TGraph[A])

11

12   def extend
13        (ctx : Context[A]): TGraph[A]
```

The Scala implementation is based on the *Graph for Scala* library developed by Peter Empen. This library provides an in-memory Graph library with a default implementation using adjacency lists and Scala inner classes. It is important to notice that although the base library can employ an underlying non-purely functional approach, the API itself is purely functional.

The library contains a generic trait `Graph[N, E]` to define graphs with nodes of type N and edges of kind E. There are four edge categories: hyperedge, directed hyperedge, undirected and directed edge. Each of these categories has predefined edge classes representing any combination of non-weighted, weighted, key-weighted, labeled and key-labeled. In our case, we will employ 3-uniform directed hypergraphs given that an edge relates three elements (origin, property and destiny). The library offers both a mutable and immutable implementation of graphs.

The functions from the *Graph for Scala* library used in this paper are given in Table 1.

Table 1. Functions employed from the *Graph for Scala* library

empty	Returns an empty Graph
nodes	List of nodes of a graph
edges	List of edges of a graph. For each edge e, we can obtain its 3 components using `e._1`, `e._2` and `e.last`
isEmpty	Checks if graph is empty
+	Adds an edge to a graph returning a new graph. A 3-edge between a, b and c is expressed as a~>b~>c

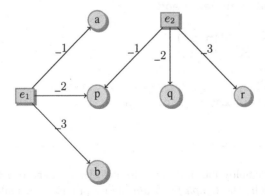

Fig. 6. RDF graph as an hypergraph. e_i are 3-hypergedges

Our implementation defines a case class `TGraphImpl` which takes a `Graph[A
,Triple]` as a parameter. `Triple` is defined as an instance of `DiHyperEdge` restricted to hyperedges of rank 3 (triples). Figure 6 depicts the graph from Figure 3 using 3-ranked hyperedges. e_1 and e_2 are the hyperedges that relate the 2 triples.

Following is a sketch of the `TGraphImpl` code:

```
 1  case class TGraphImpl[A]
 2      (graph: Graph[A,Triple])
 3      extends TGraph[A] {
 4
 5    def empty: TGraph[A] =
 6        TGraphImpl(graph.empty)
 7
 8    def nodes : Set[A] =
 9        graph.nodes.map(_.value)
10
11   def extend
12        (ctx : Context[A]): TGraph[A] = {
13   TGraphImpl(
14    ((((graph + ctx.node)
15    /: ctx.succ) {(g,p) => g +
16      Triple(ctx.node,p._1,p._2)}
17    /: ctx.pred) {(g,p) => g +
18      Triple(p._1,p._2,ctx.node)}
19      /: ctx.rels) {(g,p) => g +
20      Triple(p._1,ctx.node,p._2)})}
21
22   def decomp:
23        Option[(Context[A],TGraph[A])]={
24    if (graph.isEmpty) None
25    else {
26     val node = nodes.head
27      for {
28      pred <- pred(node)
29      succ <- succ(node)
30      rels <- rels(node)
31    } yield(Context(node,pred,succ,rels),
32          TGraphImpl(graph - node))
33   }
34  }
```

Notice that we employ the operator + to add elements and edges to a graph returning a new graph. The *Graph for Scala* library provides an implementation with mutable graphs, and another with immutable graphs. Since we work in a purely functional setting, we prefer to work with immutable data structures. Further work remains to be done to compare the efficiency of the implementations, or to further optimise an implementation.

A context of a node in a graph is defined with the following case class:

```
case class Context[A](
    node : A,
    pred: Set[(A,A)],
    succ: Set[(A,A)],
    rels: Set[(A,A)])
```

Following the encoding presented in previous section, we define RDF graphs as:

```
abstract class RDFGraph
case class Ground
    (graph : TGraph[RDFNode])
  extends RDFGraph
case class Exists
    (fn: BNode => RDFGraph)
  extends RDFGraph
```

where RDF nodes are defined by the RDFNode class.

```
abstract class RDFNode
case class IRI(iri: IRI)
  extends RDFNode
case class Literal(lit: Literal)
  extends RDFNode
case class BNode(id: String)
  extends RDFNode
```

Now, it is possible to define mapRDFGraph as:

```
def mapRDFGraph
  (fn : RDFNode => RDFNode,
   graph : RDFGraph
  ) : RDFGraph = {
  graph match {
    case Ground(g) =>
          Ground(g.mapTGraph(fn))
    case Exists(f) =>
          Exists ((x : BNode) => f(x))
  }
}
```

In the same way, we have defined other common functions like foldRDFGraph.

6 Related Work

There are a number of RDF libraries for imperative languages like Jena[6], Sesame[7] (Java), dotNetRDF[8] (C#), Redland[9] (C), RDFLib[10] (Python), RDF.rb[11] (Ruby), etc.

For dynamic languages, most of the RDF libraries are binders to some underlying imperative implementation.

banana-RDF[12] is an RDF library implementation in Scala. Although the library emphasizes type safety and immutability, the underlying implementations are Jena and Sesame.

There are some fuctional implementations of RDF libraries. Most of these employ mutable data structures. For example, scaRDF[13] started as a facade of Jena and evolved to implement the whole RDF graph machinery in Scala, employing mutable adjacency maps.

There have been several attempts to define RDF libraries in Haskell. RDF4h[14] is a complete RDF library implemented using adjacency maps, and Swish[15] provides an RDF toolkit with support for RDF inference using a Horn-style rule system. It implements some common tasks like graph merging, isomorphism and partitioning representing an RDf graph as a set of arcs.

Martin Erwig introduced the definition of inductive graphs [9]. He gives two possible implementations [8], one using version trees of functional arrays, and the other using balanced binary search trees. Both are implemented in SML. Later, Erwig implemented the second approach in Haskell which has become the FGL library.

Jeffrey and Patel-Schneider employ Agda[16] to check integrity constraints of RDF [14], and propose a programming language for the semantic web [15].

Mallea et al [16] describe the existential nature of blank nodes in RDF. Our use of existential variables was inspired by Seres and Spivey [19] and Claessen [3]. The representation is known in logic programming as 'the completion process of predicates', first described and used by Clark in 1978 [4] to deal with the semantics of negation in definite programs.

Our representation of existential variables in RDFGraphs uses a datatype with an embedded function. Fegaras and Sheard [11] describe different approaches to implement folds (also known as catamorphisms) over these kind of datatypes. Their paper contains several examples and one of them is a representation of graphs using a recursive datatype with embedded functions.

---

[6] http://jena.apache.org/
[7] http://www.openrdf.org/
[8] http://www.dotnetrdf.org/
[9] http://librdf.org/
[10] http://www.rdflib.net/
[11] http://rdf.rubyforge.org/
[12] https://github.com/w3c/banana-rdf
[13] https://code.google.com/p/scardf/
[14] http://protempore.net/rdf4h/
[15] https://bitbucket.org/doug_burke/swish
[16] https://github.com/agda/agda-web-semantic

The representation of RDF graphs using hypergraphs, and transformations between hypergraphs and bipartite graphs, have been studied by Hayes and Gutiérrez [12].

Recently, Oliveira et al. [18] define structured graphs in which sharing and cycles are represented using recursive binders, and an encoding inspired by parametric higher-order abstract syntax [2]. They apply their work to grammar analysis and transformation. It is future work to check if their approach can also be applied to represent RDF graphs.

7 Conclusions

In this paper, we have presented a simplified representation for inductive graphs that we called Inductive Triple Graphs and that can be applied to represent RDF graphs using existential variables. This representation can be implemented using immutable data structures in purely functional programming languages. A functional programming implementation makes it easier to develop basic recursion operators such as folds and maps for graphs, to obtain programs that run on multiple cores, and to prove properties about functions. We developed two different implementations: one in Haskell and another in Scala. The implementations use only standard libraries as a proof-of-concept without taking possible optimizations into account. In the future, we would like to offer a complete RDF library and to check its availability and scalability in real-time scenarios.

Acknowledgments. This work has been partially funded by Spanish project MICINN-12-TIN2011-27871 ROCAS (Reasoning on the Cloud by Applying Semantics) and by the International Excellence Campus grant of the University of Oviedo which allowed the first author to have a research stay at the University of Utrecht.

References

1. Carroll, J.J., Klyne, G.: Resource description framework (RDF): Concepts and abstract syntax. W3C recommendation, W3C (February 2004),
   http://www.w3.org/TR/2004/REC-rdf-concepts-20040210/
2. Chlipala, A.J.: Parametric higher-order abstract syntax for mechanized semantics. In: Hook, J., Thiemann, P. (eds.) Proceeding of the 13th ACM SIGPLAN International Conference on Functional Programming, ICFP 2008, Victoria, BC, Canada, September 20-28, pp. 143–156. ACM (2008)
3. Claessen, K., Ljunglöf, P.: Typed logical variables in Haskell. In: Proceedings of Haskell Workshop, Montreal, Canada, University of Nottingham, Technical Report (2000)
4. Clark, K.L.: Negation as failure. In: Gallaire, H., Minker, J. (eds.) Logic and Databases, pp. 293–322. Plenum Press (1978)
5. Cyganiak, R., Wood, D.: Resource description framework (RDF): Concepts and abstract syntax. W3C working draft, W3C (January 2013),
   http://www.w3.org/TR/rdf11-concepts/
6. Becket, D., Prud'hommeaux, E., Berners-Lee, T., Carothers, G.: Turtle, terse RDF triple language. In: World Wide Web Consortium, Working Draft, WD-Turtle (July 2012)

7. Dürst, M., Suignard, M.: Internationalized resource identifiers. Technical Report 3987, IETF (2005)
8. Erwig, M.: Fully persistent graphs - which one to choose? In: Clack, C., Hammond, K., Davie, T. (eds.) IFL 1997. LNCS, vol. 1467, pp. 123–140. Springer, Heidelberg (1998)
9. Erwig, M.: Functional programming with graphs. SIGPLAN Not. 32(8), 52–65 (1997)
10. Erwig, M.: Inductive graphs and functional graph algorithms. J. Funct. Program. 11(5), 467–492 (2001)
11. Fegaras, L., Sheard, T.: Revisiting catamorphisms over datatypes with embedded functions (or, programs from outer space). In: Proceedings of the 23rd ACM SIGPLAN-SIGACT Symposium on Principles of Programming Languages, POPL 1996, pp. 284–294. ACM, New York (1996)
12. Hayes, J., Gutierrez, C.: Bipartite graphs as intermediate model for RDF. In: McIlraith, S.A., Plexousakis, D., van Harmelen, F. (eds.) ISWC 2004. LNCS, vol. 3298, pp. 47–61. Springer, Heidelberg (2004)
13. Hughes, J.: Why Functional Programming Matters. Computer Journal 32(2), 98–107 (1989)
14. Jeffrey, A.S.A., Patel-Schneider, P.F.: Integrity constraints for linked data. In: Proc. Int. Workshop Description Logics (2011)
15. Jeffrey, A.S.A., Patel-Schneider, P.F.: As XDuce is to XML so ? is to RDF: Programming languages for the semantic web. In: Proc. Off the Beaten Track: Workshop on Underrepresented Problems for Programming Language Researchers (2012)
16. Mallea, A., Arenas, M., Hogan, A., Polleres, A.: On blank nodes. In: Aroyo, L., Welty, C., Alani, H., Taylor, J., Bernstein, A., Kagal, L., Noy, N., Blomqvist, E. (eds.) ISWC 2011, Part I. LNCS, vol. 7031, pp. 421–437. Springer, Heidelberg (2011)
17. Meijer, E., Fokkinga, M., Paterson, R., Hughes, J.: Functional Programming with Bananas, Lenses, Envelopes and Barbed Wire. In: Hughes, J. (ed.) FPCA 1991. LNCS, vol. 523, pp. 124–144. Springer, Heidelberg (1991)
18. Oliveira, B.C., Cook, W.R.: Functional programming with structured graphs. SIGPLAN Not. 47(9), 77–88 (2012)
19. Seres, S., Spivey, J.M.: Embedding Prolog into Haskell. In: Proceedings of HASKELL 1999, Department of Computer Science, University of Utrecht (1999)

A Depth-First Branch and Bound Algorithm for Learning Optimal Bayesian Networks

Brandon Malone[1] and Changhe Yuan[2]

[1] Helsinki Institute for Information Technology,
Department of Computer Science, Fin-00014 University of Helsinki, Finland
[2] Queens College/City University of New York, USA
brandon.malone@cs.helsinki.fi, changhe.yuan@qc.cuny.edu

Abstract. Early methods for learning a Bayesian network that optimizes a scoring function for a given dataset are mostly approximation algorithms such as greedy hill climbing approaches. These methods are anytime algorithms as they can be stopped anytime to produce the best solution so far. However, they cannot guarantee the quality of their solution, not even mentioning optimality. In recent years, several exact algorithms have been developed for learning optimal Bayesian network structures. Most of these algorithms only find a solution at the end of the search, so they fail to find any solution if stopped early for some reason, e.g., out of time or memory. We present a new depth-first branch and bound algorithm that finds increasingly better solutions and eventually converges to an optimal Bayesian network upon completion. The algorithm is shown to not only improve the runtime to *find* optimal network structures up to 100 times compared to some existing methods, but also prove the optimality of these solutions about 10 times faster in some cases.

1 Introduction

Early methods for learning a Bayesian network that optimizes a scoring function for a given dataset are mostly approximation algorithms such as greedy hill climbing approaches [1–3]. Even though these approximation algorithms do not guarantee optimality, they do have other nice properties. Because they usually only store the current best network plus local score information, their memory requirements are modest. They also exhibit good anytime behavior, as they can be stopped anytime to output the best solution found so far.

In recent years, several exact algorithms have been developed for learning optimal Bayesian networks, including dynamic programming [4–9], branch and bound (BB) [10], integer linear programming (ILP) [11, 12], and admissible heuristic search [13–15]. These algorithms typically do not possess inherent anytime behavior, that is, they do not find any solution before finding the optimal solution. However, some of these methods can be easily augmented with anytime behavior. For example, BB uses an approximation algorithm to find an upper bound solution in the very beginning of the search for pruning. Also, it regularly strays away from its main best-first search strategy and expands the worst nodes in the priority queue in hope to find better solutions sooner.

M. Croitoru et al. (Eds.): GKR 2013, LNAI 8323, pp. 111–122, 2014.

This paper develops an *intrinsically* anytime algorithm based on depth-first search. Several key theoretical contributions were needed to make the algorithm effective for the learning problem, including a newly formulated search graph, a new heuristic function, and a duplicate detection and repairing strategy. The algorithm combines the best characteristics of exact (guaranteed optimality when finishes) and local search (quickly identifying good solutions) algorithms. Experimentally, we show that the new algorithm has an excellent anytime behavior. The algorithm continually finds better solutions during the search; it not only *finds* the optimal solutions up to two orders of magnitude faster than some existing algorithms but also *proves* their optimality typically one order of magnitude faster when it finishes.

2 Background

This section reviews the basics of score-based Bayesian network structure learning and the shortest-path perspective of the learning problem [13], which is the basis of our new algorithm.

2.1 Learning Bayesian Network Structures

A Bayesian network is a directed acyclic graph (DAG) in which the vertices correspond to a set of random variables $\mathbf{V} = \{X_1, ..., X_n\}$, and the arcs represent dependence relations between the variables. The set of all parents of X_i are referred to as PA_i. The parameters of the network define a conditional probability distribution, $P(X_i|PA_i)$, for each X_i.

We consider the problem of learning a network structure from a dataset $\mathbf{D} = \{D_1, ..., D_N\}$, where D_i is an instantiation of all the variables in $\mathbf{V}$. A scoring function measures the goodness of fit of a network structure to $\mathbf{D}$ [2]. The problem is to find the structure which optimizes the score. In this work we used the minimum description length (MDL) [16], so it is minimization problem. Let r_i be the number of states of X_i, N_{pa_i} be the number of records in $\mathbf{D}$ consistent with $PA_i = pa_i$, and N_{x_i,pa_i} be the number of records further constraint with $X_i = x_i$. MDL is defined as follows [17].

$$score_{MDL}(G) = \sum_i score_{MDL}(X_i|PA_i), \tag{1}$$

where

$$score_{MDL}(X_i|PA_i) = H(X_i|PA_i) + \frac{\log N}{2} K(X_i|PA_i),$$

$$H(X_i|PA_i) = -\sum_{x_i,pa_i} N_{x_i,pa_i} \log \frac{N_{x_i,pa_i}}{N_{pa_i}},$$

$$K(X_i|PA_i) = (r_i - 1) \prod_{X_l \in PA_i} r_l.$$

Other decomposable scores [2] can also be used, such as BIC [18], AIC [19], BDe [20, 21] and fNML [22]. For many of those scoring functions, the best network

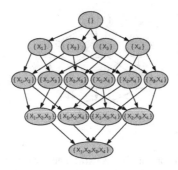

Fig. 1. A forward order graph of four variables

maximizes the score. We can multiply the scores by -1 to ensure the best network minimizes the score. For the rest of the paper, we assume the local scores, $score(X_i|PA_i)$, are computed prior to the search.

2.2 Shortest-Path Perspective

Yuan *et al.* ([13]) viewed the learning problem as a shortest-path problem in an *implicit* search graph. Figure 1 shows the search graph for four variables. The top-most node with the empty variable set is the *start* node, and the bottom-most node with the complete set is the *goal* node. An arc from **U** to **U** ∪ $\{X\}$ in the graph represents generating a successor node by adding a new variable $\{X\}$ to an existing set of variables **U**; the cost of the arc is equal to the score of the optimal parent set for X out of **U**, which is computed by considering all subsets of the variables in **U**, i.e.,

$$BestScore(X, \mathbf{U}) = \min_{PA_X \subseteq \mathbf{U}} score(X|PA_X).$$

In this search graph, each path from the start to the goal corresponds to an ordering of the variables in the order of their appearance, so we also call the graph a *forward order graph*. Each variable selects optimal parents from the variables that precede it, so combining the optimal parent sets yields an optimal structure for that ordering. The cost of a path is equal to the sum of the parent selections entailed by the ordering. The shortest path gives the global optimal structure.

2.3 Finding Optimal Parents

While searching the forward order graph, the cost of each visited arc is calculated using *parent graphs*. The parent graph for X consists of all subsets of $\mathbf{V} \setminus \{X\}$. Figure 2(a) shows a parent graph for X_1. Some subsets cannot possibly be the optimal parent set according to the following theorem [23].

Theorem 1. *Let* $\mathbf{U} \subset \mathbf{T}$ *and* $X \notin \mathbf{T}$. *If* $score(X|\mathbf{U}) < score(X|\mathbf{T})$, $\mathbf{T}$ *is not the optimal parent set for* X.

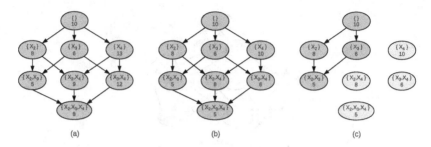

Fig. 2. A sample parent graph for variable X_1. (a) The local scores, $score(X_1, \cdot)$ for all the parent sets. (b) The optimal scores, $BestScore(X_1, \cdot)$, for each candidate parent set. (c) The unique optimal parent sets and their scores. The pruned parent sets are shown in gray. A parent set is pruned if any of its predecessors has a better score.

Table 1. Sorted scores and parent sets for X_1 after pruning sets which are not possibly optimal

$parents_{X_1}$	$\{X_2, X_3\}$	$\{X_3\}$	$\{X_2\}$	$\{\}$
$scores_{X_1}$	5	6	8	10

While we focus on the MDL score in this paper, the theorem holds for all decomposable scoring functions.

Figure 2(b) shows the optimal scores after propagating $BestScore(\cdot)$ from subsets to supersets. The score of the arc from $\mathbf{U}$ to $\mathbf{U} \cup \{X\}$ in the order graph is given by the node $\mathbf{U}$ in the parent graph for X.

The parent graph for a variable X exhaustively enumerates and stores $BestScore(X, \mathbf{U})$ for all subsets of $\mathbf{V} \setminus \{X\}$. Naively, all the parent graphs require storing $n2^{n-1}$ subsets. Due to Theorem 1, the number of *unique* optimal parent sets is often far smaller than the total number, as the same score can be optimal for many nodes. We can remove the repetitive information by only storing the optimal parent sets and their scores as sorted lists for each parent graph. The results are a set of *sparse parent graphs* [15]. Figure 2(c) shows the sparse parent graph of X_1, and Table 1 shows the corresponding sorted lists. We call these lists $scores_X$ and $parents_X$. When given a candidate parent set $\mathbf{U}$ for X, we find the optimal parents by scanning through $parents_X$ and find the first subset of $\mathbf{U}$ plus its score.

2.4   Finding the Shortest Path

Various methods can be applied to solve the shortest path problem. Dynamic programming can be considered to evaluate the order graph using a top down sweep of the graph [6, 9]. The A* algorithm in [13] evaluates the order graph with a best first search guided by an admissible heuristic function. The breadth-first branch and bound (BF-BnB) algorithm in [14] searches the order graph one layer at a time and uses an initial upper bound solution for pruning.

None of these algorithms finds a solution until the very end of the search. If they run out of resources, they yield no solution. We address these limitations by introducing a simple but very effective anytime learning algorithm.

3   A Depth-First Branch and Bound Algorithm

Our new algorithm applies *depth-first search* (DFS) to solve the shortest path problem. Depth-first search has an intrinsic anytime behavior; each time the depth-first search reaches the goal node, a new solution is found and can potentially be used to update the incumbent solution.

However, applying depth-first search to the learning problem is not as straightforward as expected. Several improvements are needed to make the algorithm effective for the particular problem, including a newly formulated search graph, a new heuristic function, and a duplicate detection and repairing strategy.

3.1   Reverse Order Graph

A key improvement comes from addressing the inefficiency in finding optimal parents in the sparse parent graphs. For a candidate parent set U for X, we need to find the optimal parents by scanning through $parents_X$ and find the first subset of U as the optimal parents. The scan is achieved by removing all the non-candidate variables $V \setminus U$ and is best implemented using bit vectors and bitwise operations. Readers are referred to [15] for more details because of the lack of space. It suffices to say here that bit vectors are used to keep track of optimal parent sets that contain allowable candidate parents, and bitwise operations are applied to the vectors to remove non-candidate variables one at a time.

For example, the first score in Table 1 is optimal for the candidate parent set $\{X_2, X_3, X_4\}$. To remove X_2 from the candidate set and find $BestScore(X_1, \{X_3, X_4\})$, we scan $parents_X$ from the beginning. The first parent set which does not include X_2 is $\{X_3\}$. To remove both X_2 and X_3 and find $BestScore(X_1, \{X_4\})$, we scan until finding a parent set which includes neither X_2 nor X_3; that is $\{\}$. In the worst case, a complete scan of the list is needed, which makes each search step expensive.

The scan can be made more efficient if only one variable needs to be removed at each search step by reusing results from previous steps. Depth-first search enables us to achieve that. We can store the bit vectors used in previous scans in the *search stack*. Since each search step in the order graph only processes one more variable, the search can start from the bit vectors on top of the stack and remove only one more variable to find the optimal parents. Also, depth-first search makes it necessary to store only at most $O(n^2)$ bit vectors at any time. Other search methods may require an exponential number of bit vectors to use the same incremental strategy.

However, the forward order graph is not very amenable to the above incremental scan, because variables are added as candidate parents as the search goes deeper. We therefore propose a newly formulated search graph. We notice that there is symmetry between the start and goal nodes in the forward order graph; the shortest path from the start node to the goal node is the same as the shortest path from the latter to the former. The reversed direction of search, however, is better because variables are removed incrementally. Therefore, we propose to use the *reverse order graph* shown in Figure 3 as our search graph. The top-most node with the complete set is the new start node, and the bottom-most empty set is the new goal node. An arc from U to $U \setminus \{X\}$

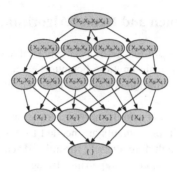

Fig. 3. A reverse order graph of four variables

represents making X a leaf node for the subnetwork over $\mathbf{U}$. The cost of the arc is $BestScore(X, \mathbf{U} \setminus \{X\})$. The task is to find the shortest path between the new start and goal nodes.

3.2 Branch and Bound

The efficiency of the depth-first search can be significantly improved by pruning. The best solution found so far is an upper bound for the optimal solution. If we can also estimate a lower bound h on the cost from the current search node to the goal, then we can calculate a lower bound on the cost of a solution using the current path to this node. If that lower bound is worse than the upper bound solution, then the current path will not lead to any better solution.

Since the reverse order graph has a different goal node from the forward order graph, we cannot use the same heuristic function in [13, 14]. We therefore design a new admissible heuristic function. At any point in the search, a set of variables remain for which we must select parents. Addtionally, we know that, at a node $\mathbf{U}$, the parent selections for the remaining variables are a subset of $\mathbf{U}$. Therefore, $BestScore(X, \mathbf{U})$ is a lower bound on the scores for each remaining variable, X. We calculate $BestScore(X, \mathbf{U})$ for the remaining X at node $\mathbf{U}$ with the bit vectors on top of the search stack. Summing over these scores gives a lower bound on the optimal subnetwork over $\mathbf{U}$. More formally, we use the following new heuristic function h^*.

Definition 1

$$h^*(\mathbf{U}) = \sum_{X \in \mathbf{U}} BestScore(X, \mathbf{U} \setminus \{X\}) \tag{2}$$

The heuristic is admissible because it allows the remaining variables to select optimal parents from among all of the other remaining variables. This relaxes the acyclic constraint on those variables. The following theorem shows the heuristic is also consistent (proof is in the appendix).

Theorem 2. *h^* is consistent.*

3.3 Duplicate Detection and Repair

A potential problem with DFS is generating duplicate nodes. A traditional DFS algorithm does not perform duplicate detection; much work can be wasted in re-expanding duplicate nodes. Our search graph contains many duplicates; a node in layer l can be generated by all its l predecessor nodes. In order to combat this problem, our algorithm stores nodes that have been expanded in a hash table and detects duplicate nodes during the search. If a better path to a generated node has already been found, we can immediately discard the duplicate.

On the other hand, we need to re-expand a node if a better path is found to it. We efficiently handle re-expansions by adopting an *iterative* DFS strategy. During each iteration, we expand each node only once. We keep a list which stores all nodes to which we find a better path. We expand those nodes in the next iteration and iterate this process until no nodes are added to the list.

We further improve the anytime behavior of the algorithm by storing the cost of the best discovered path from $\mathbf{U}$ to $goal$, $h_{exact}(\mathbf{U})$. When backtracking, we compute $h_{exact}(\mathbf{U})$ by calculating, for each successor $\mathbf{R}$, the total distance between $\mathbf{U}$ and $\mathbf{R}$ and between $\mathbf{R}$ and the goal. The minimum distance among them is $h_{exact}(\mathbf{U})$. Trivially, h_{exact} of an immediate predecessor of $goal$ is just the distance from it to $goal$. We pass this information up the call stack to calculate h_{exact} for predecessor nodes. The next time $\mathbf{U}$ is generated, we sum the distance on the current path and $h_{exact}(\mathbf{U})$. If it is better than the existing best path, $optimal$, then we update the best path found so far. We store h_{exact} in the same hash table used for duplicate detection.

3.4 Depth-First Search

Algorithm 1 (at the end of the paper) provides the pseudocode for the DFS algorithm. After initializing the data structures (lines 25 - 27), the EXPAND function is called with the start node as input. At each node $\mathbf{U}$, we make one variable X as a leaf (line 5) and select its optimal parents from among $\mathbf{U}$ (lines 6, 7). We then check if that is the best path to the subnetwork $\mathbf{U} \setminus \{X\}$ (lines 10 - 12). The bitwise operators are used to remove X as a possible parent for the remaining variables (lines 13 - 15). Parents are recursively selected for the remaining variables (line 16). Because we did not modify $valid$, the call stack maintains the valid parents before removing X, so we perform the bitwise operators for the next leaf. After trying every remaining variable as a leaf, we check if a better path to the goal has been found (line 22). During each iteration of the search, we keep a list that tracks nodes to which we find a better path (line 10) and repair those nodes in the next iteration by iteratively performing the DFS (lines 29 - 37). The search continues until no nodes are added to the list.

The depth-first nature of the algorithm means we typically find a path to the goal after n node expansions; it can be stopped at any time to return the best path found so far. For conciseness, Algorithm 1 only includes the main logic in computing the optimal score; we use the network reconstruction technique described in [14].

Algorithm 1. A DFBnB Search Algorithm

 1: **procedure** EXPAND($\mathbf{U}, valid, toRepair$)
 2:    **for each** $X \in \mathbf{U}$ **do**
 3:        $BestScore(X, \mathbf{U} \setminus \{X\}) \leftarrow scores_X[firstSetBit(valid_X)]$
 4:        $g \leftarrow g(\mathbf{U}) + BestScore(X, \mathbf{U} \setminus \{X\})$
 5:        $duplicate \leftarrow exists(g(\mathbf{U} \setminus \{X\}))$
 6:        **if** $g < g(\mathbf{U} \setminus \{X\})$ **then** $g(\mathbf{U} \setminus \{X\}) \leftarrow g$
 7:        **if** $duplicate$ and $g < g(\mathbf{U} \setminus \{X\})$ **then** $toRepair \leftarrow toRepair \cup \{\mathbf{U}, g\}$
 8:        $f \leftarrow h(\mathbf{U} \setminus \{X\}) + g(\mathbf{U} \setminus \{X\})$
 9:        **if** $!duplicate$ and $f < optimal$ **then**
10:            **for each** $Y \in \mathbf{U}$ **do**
11:                $valid'_Y \leftarrow valid_Y \& \sim parents_Y(X)$
12:            **end for**
13:            $expand(\mathbf{U} \setminus \{X\}, valid')$
14:        **end if**
15:        **if** $h_{exact}(\mathbf{U}) > BestScore(X, \mathbf{U} \setminus \{X\}) + h_{exact}(\mathbf{U} \setminus \{X\})$ **then**
16:            $h_{exact}(\mathbf{U}) \leftarrow BestScore(X, \mathbf{U} \setminus \{X\}) + h_{exact}(\mathbf{U} \setminus \{X\})$
17:        **end if**
18:    **end for**
19:    **if** $optimal > h_{exact}(\mathbf{U}) + g(\mathbf{U})$ **then** $optimal \leftarrow h_{exact}(\mathbf{U}) + g(\mathbf{U})$
20: **end procedure**

21: **procedure** MAIN($\mathbf{D}$)
22:    **for each** $X \in \mathbf{V}$ **do**
23:        $scores_X, parents_X \leftarrow initDataStructures(X, \mathbf{D})$
24:    **end for**
25:    $h_{exact}(\mathbf{U}) \leftarrow 0$
26:    $toRepair_l \leftarrow \{\mathbf{V}, 0\}$
27:    **while** $|toRepair_l| > 0$ **do**
28:        **for each** $\{\mathbf{V}, g\} \in toRepair_l$ **do**
29:            **if** $g(\mathbf{V}) > g$ **then**
30:                $g(\mathbf{V}) \leftarrow g$
31:                $expand(\mathbf{V}, valid, toRepair_{l+1})$
32:            **end if**
33:        **end for**
34:        $toRepair_l \leftarrow toRepair_{l+1}$
35:    **end while**
36: **end procedure**

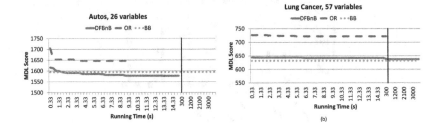

Fig. 4. The scores of the best networks learned by DFBnB, OR and BB on two datasets over time. The algorithms are given up to one hour (3600 s) of runtime. A black bar indicates an increase in the time scale. OR reached a local optimum and terminated within 300s. DFBnB found and proved the optimal solution for the Autos dataset within the time limit. BB did not improve its initial solution or find the optimal solution within the time limit for either dataset.

4 Experiments

We tested the DFBnB algorithm against several existing structure learning algorithms: dynamic programming (DP) [6], breadth-first branch and bound (BFBnB) [14], branch and bound (BB) [10], and optimal reinsertion (OR) [3]. The source code for DFBnB is available online (http://url.cs.qc.cuny.edu/software/URLearning.html). Experiments were performed on a PC with 3.07 GHz Intel i7, 16 GB RAM and 500 GB hard disk. We used benchmark datasets from UCI [24]. Records with missing values were removed. Variables were discretized into two states.

4.1 Comparison of Anytime Behavior

We first compared the anytime behavior of DFBnB to those of BB and OR on two datasets of up to 57 variables. Other datasets that we tested show similar results. We chose to compare to BB because, like DFBnB, it has anytime behavior and optimality guarantees. We compare to OR as a representative for local search techniques because several studies [23, 25] suggest that it performs well. We ran each algorithm for one hour and plotted the scores of the networks learned by each algorithm as a function of time in Figure 4.

As the convergence curves of these algorithms show, OR was always the first algorithm to terminate; it reached local optima far from optimal and was unable to escape. BB did not finish within the time limit on either dataset. The convergence curves of BB stayed flat for both. BB used a sophisticated approximation algorithm to find its initial solution. That means BB was not able to improve its initial solutions in an hour of runtime, so the true anytime behavior of BB is unclear from the results. In comparison, DFBnB intrinsically finds all solutions, so its curves provide a reliable indication of its anytime behavior. On both datasets, DFBnB continually found better solutions during the search, and was able to find and prove the optimality of its solution on the Autos dataset within 5 minutes. BB learned a network whose score was 0.7% lower than DFBnB on the Lung Cancer dataset.

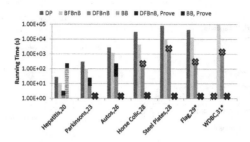

Fig. 5. A comparison on the running time for BFBnB, DP, DFBnB and BB. DP and BFBnB were given an arbitrary amount of time to complete because of their efficient use of external memory. DFBnB and BB were given a maximum running time of one hour. For DP, an "X" for a dataset means the external memory usage exceeded 500 GB. For DFBnB and BB, an "X" means it did not prove the optimality of the solution found. An "X" at time 0 for BB means that BB did not improve its initial greedy solution and did not find the optimal solution. DFBnB found the optimal solution, though may not have proved it, for all datasets which do not have an "*" beside the name.

4.2   Comparison of Running Time

We compared the running time of DFBnB to those of DP, BFBnB and BB. DFBnB and BB were again given a one hour limit. We let BFBnB and DP run longer in order to obtain the optimal solutions for evaluation. For this comparison, we considered both the time to *find* the best structure and to *prove* its optimality by DFBnB and BB. The results in Figure 5 demonstrate that DFBnB finds the optimal solution nearly two orders of magnitude faster than the other algorithms, even though it was not always able to prove optimality due to lack of RAM. When it had enough RAM, though, DFBnB proved optimality nearly an order of magnitude faster that other algorithms.

The latter result is somewhat surprising given that BFBnB is based a similar shortest-path formulation and was shown to be as efficient as A* [14]. The impressive performance of DFBnB comes from the incremental scan technique used to find optimal parents in the sparse parent graphs. In comparison, each step of BFBnB and A* starts from scratch and removes all non-candidate variables to find optimal parents.

However, DFBnB does not take advantage of disk. The program stops when its hash table fills RAM, so it is unable to prove the optimality for some of the searches. Nevertheless, the search finds optimal solutions for all but the largest two datasets (verified with BFBnB). BB proved the optimality of its solution on only the smallest dataset. On all other cases, it did not improve its initial solutions. DFBnB found better solutions than BB on all these datasets. Note Lung Cancer is not included here because none of the algorithms proved the optimality of their solutions within the resource limits.

5   Conclusion

In this paper we have proposed an anytime algorithm for learning exact Bayesian network structures. The algorithm introduces a new shortest-path formulation of the problem to use depth-first search. We introduced the reverse order graph and a new heuristic function to improve the efficiency of the search algorithm. A duplicate detection and repair strategy is also used to avoid unnecessary re-expansions while maintaining excellent anytime behavior.

Experimental results demonstrated improved anytime behavior of our algorithm compared to some other structure learning algorithms. Our algorithm often finds good solutions more quickly than the other anytime algorithms, even though the local search techniques forgo optimality guarantees. Optimal reinsertion is one of the best performing approximation methods, but our results show that it often finds solutions that are far from optimal.

Our algorithm is also guaranteed to converge to an optimal solution when it has enough RAM. In those cases, it proves the optimality of its solution almost an order of magnitude faster than other search algorithms. Consequently, our algorithm combines the best qualities of both globally optimal algorithms and local search algorithms.

6 Appendix

Proof of Theorem 2.

Proof. For any successor node $\mathbf{R}$ of $\mathbf{U}$, let $Y \in \mathbf{U} \setminus \mathbf{R}$, let $c(\mathbf{U}, \mathbf{R})$ be the cost of using Y as the leaf of $\mathbf{U}$. We have

$$
\begin{aligned}
h^*(\mathbf{U}) &= \sum_{X \in \mathbf{U}} BestScore(X, \mathbf{U} \setminus \{X\}) \\
&\leq \sum_{X \in \mathbf{U}, X \neq Y} BestScore(X, \mathbf{U}) + BestScore(Y, \mathbf{U}) \\
&\leq \sum_{X \in \mathbf{R}} BestScore(X, \mathbf{R} \setminus \{X\}) + BestScore(Y, \mathbf{U} \setminus \{Y\}) \\
&= h^*(\mathbf{R}) + c(\mathbf{U}, \mathbf{R}).
\end{aligned}
$$

The inequality holds because the variables in $\mathbf{R}$ have fewer parents to choose from after making Y a leaf. Hence, h^* is consistent.

Acknowledgements. This work was supported by the Academy of Finland (Finnish Centre of Excellence in Computational Inference Research COIN, 251170) and NSF grants IIS-0953723 and IIS-1219114.

References

1. Cooper, G.F., Herskovits, E.: A Bayesian method for the induction of probabilistic networks from data. Machine Learning 9, 309–347 (1992)
2. Heckerman, D.: A tutorial on learning with Bayesian networks. In: Jordan, M. (ed.) Learning in Graphical Models. NATO ASI Series, vol. 89, pp. 301–354. Springer, Netherlands (1998)
3. Moore, A., Wong, W.K.: Optimal reinsertion: A new search operator for accelerated and more accurate Bayesian network structure learning. In: Proceedings of the International Conference on Machine Learning, pp. 552–559 (2003)

4. Koivisto, M., Sood, K.: Exact Bayesian structure discovery in Bayesian networks. Journal of Machine Learning Research, 549–573 (2004)
5. Ott, S., Imoto, S., Miyano, S.: Finding optimal models for small gene networks. In: Pacific Symposium on Biocomputing, pp. 557–567 (2004)
6. Silander, T., Myllymaki, P.: A simple approach for finding the globally optimal Bayesian network structure. In: Proceedings of the 22nd Conference on Uncertainty in Artificial Intelligence. AUAI Press, Arlington (2006)
7. Singh, A., Moore, A.: Finding optimal Bayesian networks by dynamic programming. Technical report, Carnegie Mellon University (June 2005)
8. Parviainen, P., Koivisto, M.: Exact structure discovery in Bayesian networks with less space. In: Proceedings of the 25th Conference on Uncertainty in Artificial Intelligence (2009)
9. Malone, B., Yuan, C., Hansen, E.: Memory-efficient dynamic programming for learning optimal Bayesian networks. In: Proceedings of the 25th National Conference on AI (2011)
10. de Campos, C.P., Ji, Q.: Efficient learning of Bayesian networks using constraints. Journal of Machine Learning Research 12, 663–689 (2011)
11. Cussens, J.: Bayesian network learning with cutting planes. In: Proceedings of the 27th Conference on Uncertainty in Artificial Intelligence, pp. 153–160. AUAI Press (2011)
12. Jaakkola, T., Sontag, D., Globerson, A., Meila, M.: Learning Bayesian network structure using LP relaxations. In: Proceedings of the 13th International Conference on Artificial Intelligence and Statistics (2010)
13. Yuan, C., Malone, B., Wu, X.: Learning optimal Bayesian networks using A* search. In: Proceedings of the 22nd International Joint Conference on Artificial Intelligence (2011)
14. Malone, B., Yuan, C., Hansen, E., Bridges, S.: Improving the scalability of optimal Bayesian network learning with external-memory frontier breadth-first branch and bound search. In: Proceedings of the 27th Annual Conference on Uncertainty in Artificial Intelligence (2011)
15. Yuan, C., Malone, B.: An improved admissible heuristic for finding optimal Bayesian networks. In: Proceedings of the 28th Conference on Uncertainty in AI (2012)
16. Lam, W., Bacchus, F.: Learning Bayesian belief networks: An approach based on the MDL principle. Computational Intelligence 10, 269–293 (1994)
17. Tian, J.: A branch-and-bound algorithm for MDL learning Bayesian networks. In: Proceedings of the 16th Conference on Uncertainty in Artificial Intelligence, pp. 580–588. Morgan Kaufmann Publishers Inc. (2000)
18. Schwarz, G.: Estimating the dimension of a model. The Annals of Statistics 6, 461–464 (1978)
19. Akaike, H.: Information theory and an extension of the maximum likelihood principle. In: Proceedings of the Second International Symposium on Information Theory, pp. 267–281 (1973)
20. Buntine, W.: Theory refinement on Bayesian networks. In: Proceedings of the Seventh Conference on Uncertainty in Artificial Intelligence, pp. 52–60. Morgan Kaufmann Publishers Inc., San Francisco (1991)
21. Heckerman, D., Geiger, D., Chickering, D.M.: Learning Bayesian networks: The combination of knowledge and statistical data. Machine Learning 20, 197–243 (1995)
22. Silander, T., Roos, T., Kontkanen, P., Myllymaki, P.: Factorized normalized maximum likelihood criterion for learning Bayesian network structures. In: Proceedings of the 4th European Workshop on Probabilistic Graphical Models (PGM 2008), pp. 257–272 (2008)
23. Teyssier, M., Koller, D.: Ordering-based search: A simple and effective algorithm for learning Bayesian networks. In: Proceedings of the Twenty-First Conference on Uncertainty in Artificial Intelligence, pp. 584–590. AUAI Press, Arlington (2005)
24. Frank, A., Asuncion, A.: UCI machine learning repository (2010)
25. Tsamardinos, I., Brown, L., Aliferis, C.: The max-min hill-climbing Bayesian network structure learning algorithm. Machine Learning 65, 31–78 (2006), 10.1007/s10994-006-6889-7

Learning Bayes Nets for Relational Data with Link Uncertainty

Zhensong Qian and Oliver Schulte*

School of Computing Science
Simon Fraser University
Vancouver-Burnaby, Canada
{zqian,oschulte}@sfu.ca
http://www.cs.sfu.ca/~oschulte/

Abstract. We present an algorithm for learning correlations among link types and node attributes in relational data that represent complex networks. The link correlations are represented in a Bayes net structure. This provides a succinct graphical way to display relational statistical patterns and support powerful probabilistic inferences. The current state of the art algorithm for learning relational Bayes nets captures only correlations among entity attributes *given* the existence of links among entities. The models described in this paper capture a wider class of correlations that involve uncertainty about the link structure. Our base line method learns a Bayes net from join tables directly. This is a statistically powerful procedure that finds many correlations, but does not scale well to larger datasets. We compare join table search with a hierarchical search strategy.

1   Introduction

Scalable link analysis for relational data with multiple link types is a challenging problem in network science. We describe a method for learning a Bayes net that captures simultaneously correlations between link types, link features, and attributes of nodes. Such a Bayes net provides a succinct graphical representation of complex statistical-relational patterns. A Bayes net model supports powerful probabilistic reasoning for answering "what-if" queries about the probabilities of uncertain outcomes conditional on observed events. Previous work on learning Bayes nets for relational data was restricted to correlations among attributes given the existence of links [16]. The larger class of correlations examined in our new algorithms includes two additional kinds:

1. Dependencies between different types of links.
2. Dependencies among node attributes given the *absence* of a link between the nodes.

* This research was supported by a Discovery Grant to Oliver Schulte from the Canadian Natural Sciences and Engineering Council. And Zhensong Qian was also supported by a grant from the China Scholarship Council. The preliminary version of this paper was presented in the IJCAI 2013 GKR workshop.

M. Croitoru et al. (Eds.): GKR 2013, LNAI 8323, pp. 123–137, 2014.
© Springer International Publishing Switzerland 2014

Discovering such dependencies is useful for several applications.

Knowledge Discovery. Dependencies provide valuable insights in themselves. For instance, a web search manager may wish to know whether if a user searches for a video in Youtube for a product, they are also likely to search for it on the web.

Relevance Determination. Once dependencies have been established, they can be used as a relevance filter for focusing further network analysis only on statistically significant associations. For example, the classification and clustering methods of Sun and Han [19] for heterogeneous networks assume that a set of "metapaths" have been found that connect link types that are associated with each other.

Query Optimization. The Bayes net model can also be used to estimate relational statistics, the frequency with which statistical patterns occur in the database [17]. This kind of statistical model can be applied for database query optimization [4].

Approach. We consider three approaches to multiple link analysis with Bayes nets.

Flat Search. Applies a standard Bayes net learner to a single large join table. This table is formed as follows: (1) take the cross product of entity tables. (An entity table lists the set of nodes of a given type.) (2) For each tuple of entities, add a relationship indicator whose value "true" or "false" indicates whether the relationship holds among the entities.

Hierarchical Search. Conducts bottom-up search through the lattice of table joins hierarchically. Dependencies (Bayes net edges) discovered on smaller joins are propagated to larger joins. The different table joins include information about the presence or absence of relationships as in the flat search above. This is an extension of the current state of the art Bayes net learning algorithm for relational data [16].

Evaluation. We compare the learned models using standard scores (e.g., Bayes Information Criterion, log-likelihood). These results indicate that both flat search and hierarchical search are effective at finding correlations among link types. Flat search can on some datasets achieve a higher score by exploiting attribute correlations that depend on the absence of relationships. Structure learning time results indicate that hierarchical search is substantially more scalable.

The main contribution of this paper is extending the current state-of-the-art Bayes net learner to model correlations among different types of links, with a comparison of a flat and a hierarchical search strategy.

Paper Organization. We describe Bayes net models for relational data (Poole's Parametrized Bayes Nets). Then we present the learning algorithms, first flat search then hierarchical search. We compare the models on four databases from different domains.

2   Related Work

Approaches to structure learning for directed graphical models with link uncertainty have been previously described, such as [3]. However to our knowledge, no implementations of such structure learning algorithms for directed graphical models are available. Our system builds on the state-of-the-art Bayes net learner for relational data, whose code is available at [6]. Implementations exist for other types of graphical models, specifically Markov random fields (undirected models) [2] and dependency networks (directed edges with cycles allowed) [10]. Structure learning programs for Markov random fields are provided by Alchemy [2] and Khot et al [9]. Khot et al. use boosting to provide a state-of-the-art dependency network learner. None of these programs are able to return a result on half of our datasets because they are too large. For space reasons we restrict the scope of this paper to directed graphical models and do not go further into undirected model. For an extensive comparison of the learn-and-join Bayes net learning algorithm with Alchemy please see [16].

3   Background and Notation

Poole introduced the Parametrized Bayes net (PBN) formalism that combines Bayes nets with logical syntax for expressing relational concepts [12]. We adopt the PBN formalism, following Poole's presentation.

3.1   Bayes Nets for Relational Data

A **population** is a set of individuals. Individuals are denoted by lower case expressions (e.g., *bob*). A **population variable** is capitalized. A **functor** represents a mapping $f : \mathcal{P}_1, \ldots, \mathcal{P}_a \to V_f$ where f is the name of the functor, each $\mathcal{P}_i$ is a population, and V_f is the output type or **range** of the functor. In this paper we consider only functors with a finite range, disjoint from all populations. If $V_f = \{T, F\}$, the functor f is a (Boolean) **predicate**. A predicate with more than one argument is called a **relationship**; other functors are called **attributes**. We use uppercase for predicates and lowercase for other functors.

A **Bayes Net (BN)** is a directed acyclic graph (DAG) whose nodes comprise a set of random variables and conditional probability parameters. For each assignment of values to the nodes, the joint probability is specified by the product of the conditional probabilities, $P(child|parent\_values)$. A **Parametrized random variable** is of the form $f(X_1, \ldots, X_a)$, where the populations associated with the variables are of the appropriate type for the functor. A **Parametrized Bayes Net** (PBN) is a Bayes net whose nodes are Parametrized random variables [12]. If a Parametrized random variable appears in a Bayes net, we often refer to it simply as a node.

Table 1. A relational schema for a university domain. Key fields are underlined. An instance for this schema is given in Figure 1.

> *Student*(*student_id*, *intelligence*, *ranking*)
> *Course*(*course_id*, *difficulty*, *rating*)
> *Professor* (*professor_id*, *teaching_ability*, *popularity*)
> *Registered* (*student_id*, *Course_id*, *grade*, *satisfaction*)
> *Teaches*(*professor_id*, *course_id*)

3.2 Databases and Table Joins

We begin with a standard **relational schema** containing a set of tables, each with key fields, descriptive attributes, and possibly foreign key pointers. A **database instance** specifies the tuples contained in the tables of a given database schema. We assume that tables in the relational schema can be divided into *entity tables* and *relationship tables*. This is the case whenever a relational schema is derived from an entity-relationship model (ER model) [21, Ch.2.2]. The functor formalism is rich enough to represent the constraints of an ER schema by the following translation: Entity sets correspond to types, descriptive attributes to functions, relationship tables to predicates, and foreign key constraints to type constraints on the arguments of relationship predicates. Assuming an ER design, a relational structure can be visualized as a complex network [13, Ch.8.2.1]: individuals are nodes, attributes of individuals are node labels, relationships correspond to (hyper)edges, and attributes of relationships are edge labels. Conversely, a complex network can be represented using a relational database schema.

Table 1 shows a relational schema for a database related to a university. In this example, there are two entity tables: a *Student* table and a *Course* table. There is one relationship table *Registered* with foreign key pointers to the *Student* and *Course* tables whose tuples indicate which students have registered in which courses. Figure 1 displays a small database instance for this schema together with a Parametrized Bayes Net (omitting the *Teaches* relationship for simplicity.)

The **natural table join**, or simply join, of two or more tables contains the rows in the Cartesian products of the tables whose values match on common fields. In logical terms, a join corresponds to a conjunction [21].

4 Bayes Net Learning With Link Correlation Analysis

We outline the two methods we compare in this paper, flat search and hierarchical search.

4.1 Flat Search

The basic idea for flat search is to apply a standard propositional or single-table Bayes net learner to a single large join table. To learn correlations between link

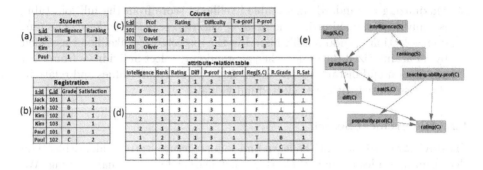

Fig. 1. Database Table Instances: (a) *Student* (b) *Registered* (c) *Course*. To simplify, we added the information about professors to the courses that they teach. (d) The attribute-relation table *Registered*⁺ derived from *Registered*, which lists for each pair of entities their descriptive attributes, whether they are linked by *Registered*, and the attributes of a link if it exists. (e) A Parametrized Bayes Net for the university schema.

types, we need to provide the Bayes net with data about when links are present *and* when they are absent. To accomplish this, we add to each relationship table a **link indicator column**. This columns contains T if the link is present between two entities, and F if the link is absent. (The entities are specified in the primary key fields.) We add rows for all pairs of entities of the right type for the link, and enter T or F in the link indicator column depending on whether a link exists or not. We refer to relationship tables with a link indicator column as **extended** tables. Extended tables are readily computed using SQL queries. If we omit the entity Ids from an extended table, we obtain the **attribute-relation** table that lists (1) all attributes for the entities involved, (2) whether a relationship exists and (3) the attributes of the relationship if it exists. If the attribute-relation table is derived from a relationship R, we refer to it as R^{+}.

The attribute-relation table is readily defined for a set of relationships: take the cross-product of all populations involved, and add a link indicator column for each relationship in the set. For instance, if we wanted to examine correlations that involve both the *Registered* and the *Teaches* relationships, we would form the cross-product of the entity types *Student, Course, Professor* and build an attribute-relation table that contains two link indicator columns $Registered(S, C)$ and $Teaches(P, C)$. The **full join table** is the attribute-relation table for all relationships in the database.

The **flat search Bayes net learner** takes a standard Bayes net learner and applies it to the full join table to obtain a single Parametrized Bayes net. The results of [14] can be used to provide a theoretical justification for this procedure; we outline two key points.

1. The full join table correctly represents the *sufficient statistics*[5,14] of the database: using the full join table to compute the frequency of a joint value assignment for Parametrized Random Variables is equivalent to the frequency with which this assignment holds in the database.

2. Maximizing a standard single-table likelihood score from the full join table
   is equivalent to maximizing the *random selection pseudo likelihood*. The random selection pseudo log-likelihood is the expected log-likelihood assigned
   by a Parametrized Bayes net when we randomly select individuals from each
   population and instantiate the Bayes net with attribute values and relationships associated with the selected individuals.

4.2   Hierarchical Search

Khosravi *et al.* [16] present the learn-and-join structure learning algorithm. The
algorithm upgrades a single-table Bayes net learner for relational learning. We
describe the fundamental ideas of the algorithm; for further details please see
[16].

The key idea is to build a Bayes net for the entire database by level-wise search
through the *table join lattice*. The user chooses a single-table Bayes net learner.
The learner is applied to table joins of size 1, that is, regular data tables. Then
the learner is applied to table joins of size $s, s + 1, \ldots$, with the constraint that
larger join tables inherit the absence or presence of learned edges from smaller
join tables. These edge constraints are implemented by keeping a global cache of
forbidden and required edges. Algorithm 1 provides pseudocode for the previous
learn-and-join algorithm (LAJ) [15].

To extend the learn-and-join algorithm for multiple link analysis, we replace
the natural join in line 7 by the extended join (more precisely, by the attribute-
relation tables derived from the extended join). The natural join contains only
tuples that appear in all relationship tables. Compared to the extended join, this
corresponds to considering only rows where the link indicator columns have the
value T. When the propositional Bayes net learner is applied to such a table, the
link indicator variable appears like a constant. Therefore the BN learner cannot
find any correlations between the link indicator variable and other nodes, nor can
it find correlations among attributes conditional on the link indicator variable
being F. Thus the previous LAJ algorithm finds only correlations between entity
attributes conditional on the existence of a relationship. In sum, hierarchical
search with link correlations can be described as follows.

1. Run the previous LAJ algorithm (Algorithm 1) using natural joins.
2. Starting with the constraints from step 1, run the LAJ algorithm where
   extended joins replace natural joins. That is, for each relationship set shown
   in the lattice of Figure 2, apply the single-table Bayes net learner to the
   extended join for the relationship set.

5   Evaluation

All experiments were done on a QUAD CPU Q6700 with a 2.66GHz CPU and
8GB of RAM. The LAJ code and datasets are available on the world-wide web
[6]. We made use of the following single-table Bayes Net search implementation:
GES search [1] with the BDeu score as implemented in version 4.3.9-0 of CMU's
Tetrad package (structure prior uniform, ESS=10; [20]).

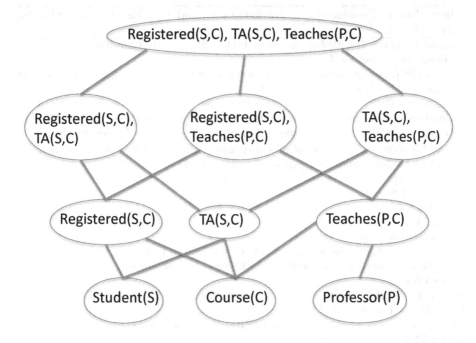

Fig. 2. A lattice of relationship sets for the university schema of Table 1. Links from entity tables to relationship tables correspond to foreign key pointers.

Methods Compared We compared the following methods.

LAJ The previous LAJ method without link correlations (Algorithm 1).
LAJ+ The new LAJ method that has the potential to find link correlations (Algorithm 1 with the extended join tables instead of natural join tables).
Flat Applies the single-table Bayes net learner to the full join table.

To implement Flat Search and the LAJ+ algorithm efficiently, we apply the Fast Möbius Transform to compute tables of sufficient statistics that involve negated relationships. We discuss the details further in Section 6.

Performance Metrics We report learning time, log-likelihood, Bayes Information Criterion (BIC), and the Akaike Information Criterion (AIC). BIC and AIC are standard scores for Bayes nets [1], defined as follows. We write

$$L(\hat{G}, \mathbf{d})$$

for the log-likelihood score, where $\hat{G}$ is the BN G with its parameters instantiated to be the maximum likelihood estimates given the dataset $\mathbf{d}$, and the quantity $L(\hat{G}, \mathbf{d})$ is the log-likelihood of $\hat{G}$ on $\mathbf{d}$.

Algorithm 1. Pseudocode for previous Learn-and-Join Structure Learning for Lattice Search

---

Input: Database $\mathcal{D}$ with $E_1,..E_e$ entity tables, $R_1,...R_r$ Relationship tables,
Output: Bayes Net for $\mathcal{D}$
Calls: PBN: Any propositional Bayes net learner that accepts edge constraints and a single table of cases as input.
Notation: PBN(T, Econstraints) denotes the output DAG of PBN. Get-Constraints(G) specifies a new set of edge constraints, namely that all edges in G are required, and edges missing between variables in G are forbidden.

1: Add descriptive attributes of all entity and relationship tables as variables to G. Add a boolean indicator for each relationship table to G.
2: Econstraints = $\emptyset$ [Required and Forbidden edges]
3: **for** m=1 to e **do**
4:     Econstraints += Get-Constraints(PBN(E_m , $\emptyset$))
5: **end for**
6: **for** m=1 to r **do**
7:     N_m := natural join of R_m and entity tables linked to R_m
8:     Econstraints += Get-Constraints(PBN(N_m, Econstraints))
9: **end for**
10: **for all** N_i and N_j with a foreign key in common **do**
11:     K_{ij} := join of N_i and N_j
12:     Econstraints += Get-Constraints(PBN(K_{ij}, Econstraints))
13: **end for**
14: **return** Bayes Net defined by Econstraints.

The BIC score is defined as follows [1,14]

$$BIC(G, \mathbf{d}) = L(\hat{G}, \mathbf{d}) - par(G)/2 \times ln(m)$$

where the data table size is denoted by m, and $par(G)$ is the number of free parameters in the structure G. The AIC score is given by

$$AIC(G, \mathbf{d}) = L(\hat{G}, \mathbf{d}) - par(G).$$

Selection by AIC is asympotically equivalent to selection by cross-validation, so we may view it as a closed-form approximation to cross-validation, which is computationally demanding for relational datasets.

Datasets. We used one synthetic and three benchmark real-world databases, with the modifications described by Schulte and Khosravi [16]. See that article for more details.

University Database. We manually created a small dataset, based on the schema given in Table 1. The dataset is small and is used as a testbed for the correctness of our algorithms.

Table 2. Size of datasets in total number of table tuples

Dataset	#tuples
University	662
Movielens	1585385
Mutagenesis	1815488
Hepatitis	2965919
Small-Hepatitis	19827

MovieLens Database. A dataset from the UC Irvine machine learning repository. The data are organized in 3 tables (2 entity tables, 1 relationship table, and 7 descriptive attributes).

Mutagenesis Database. A dataset widely used in ILP research. It contains two entity tables and two relationships.

Hepatitis Database. A modified version of the PKDD'02 Discovery Challenge database. The data are organized in 7 tables (4 entity tables, 3 relationship tables and 16 descriptive attributes). In order to make the learning feasible, we undersampled Hepatitis database to keep the ratio of positive and negative link indicator equal to one.

Table 3. Data Preprocessing: Table Join Time in seconds

Dataset	Data Processing Time
University	1.205
Movielens	1.539
Mutagenesis	0.723
Small-Hepatitis	57.794

Table 4. Model Structure Learning Time in seconds

Dataset	Flat	LAJ+	LAJ
University	1.916	1.183	0.291
Movielens	38.767	18.204	1.769
Mutagenesis	3.231	3.448	0.982
Small-Hepatitis	9429.884	7.949	10.617

5.1   Results

Learning Times. Table 3 shows the data preprocessing time that the different methods require for table joins. This is the same for all methods, namely the cost of computing the full join table using the fast Möbius transform described in Section 6. Table 4 provides the model search time for each of the link analysis methods. On the smaller and simpler datasets, all search strategies are fast, but on the medium-size and more complex datasets (Hepatitis, MovieLens), hierarchical search is much faster due to its use of constraints.

Table 5. Statistical Performance of different Searching Algorithms by dataset

University	BIC	AIC	log-likelihood	# Parameter
Flat	-17638.27	-12496.72	-10702.72	1767
LAJ+	-13495.34	-11540.75	-10858.75	655
LAJ	-13043.17	-11469.75	-10920.75	522

MovieLens	BIC	AIC	log-likelihood	# Parameter
Flat	-4912286.87	-4911176.01	-4910995.01	169
LAJ+	-4911339.74	-4910320.94	-4910154.94	154
LAJ	-4911339.74	-4910320.94	-4910154.94	154

Mutagenesis	BIC	AIC	log-likelihood	# Parameter
Flat	-21844.67	-17481.03	-16155.03	1289
LAJ+	-47185.43	-28480.33	-22796.33	5647
LAJ	-30534.26	-25890.89	-24479.89	1374

Hepatitis	BIC	AIC	log-likelihood	# Parameter
Flat	-7334391.72	-1667015.81	-301600.81	1365357
LAJ+	-457594.18	-447740.51	-445366.51	2316
LAJ	-461802.76	-452306.05	-450018.05	2230

Statistical Scores. As expected, adding edges between link nodes improves the statistical data fit: the link analysis methods LAJ+ and Flat perform better than the learn-and-join baseline in terms of log-likelihood on all datasets shown in table 5, except for MovieLens where the Flat search has a lower likelihood. On the small synthetic dataset University, flat search appears to overfit whereas the hierarchical search methods are very close. On the medium-sized dataset MovieLens, which has a simple structure, all three methods score similarly. Hierarchical search finds no new edges involving the single link indicator node (i.e., LAJ and LAJ+ return the same model).

The most complex dataset, Hepatitis, is a challenge for flat search, which seems to overfit severely with a huge number of parameters that result in a model selection score that is an order of magnitude worse than for hierarchical search. Because of the complex structure of the Hepatitis schema, the hierarchy constraints appear to be effective in combating overfitting.

The situation is reversed on the Mutagenesis dataset where flat search does well: compared to previous LAJ algorithm, it manages to fit the data better with a less complex model. Hierachical search performs very poorly compared to flat

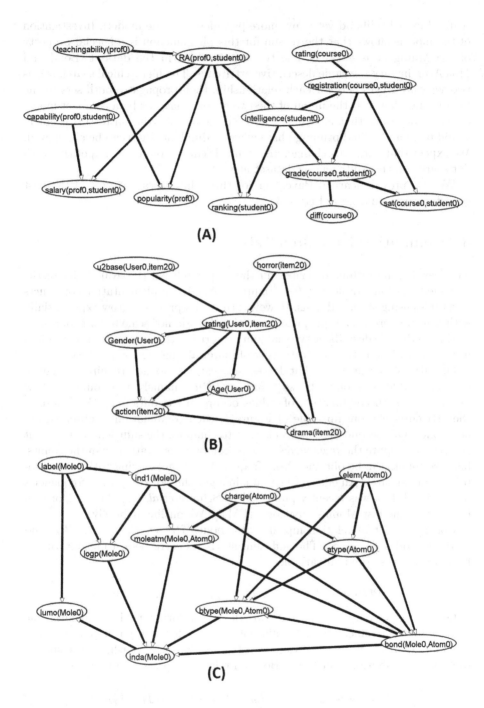

Fig. 3. Learned Parametrized Bayes Net for 3 complete datasets: (A) University Database, (B) MovieLens Database, (C) Mutagenesis Database

search (lower likelihood yet many more parameters in the model). Investigation of the models shows that the reason for this phenomenon is a special property of the Mutagenesis dataset: The two relationships in the dataset, Bond and MoleAtm, involve the same descriptive attributes. The hierarchical search learns two separate Bayes nets for each relationship, then propagates both sets to the final result. However, the union of the two graphs may not be a compact model of the associations that hold in the entire database. A solution to this problem would be to add a final pruning phase where redundant edges can be eliminated. We expect that with this change, hierarchical search would be competitive with flat search on the Mutagenesis dataset as well.

We also provide learned Bayes net for three databases in terms of the best statistical performance in Figure 3.

6   Computing Data Join Tables

The learning algorithms described in this paper rely on the availability of the extended relational tables R^+ (see Figure 1). A naive implementation constructs this tables using standard joins. However, the cross-products grow exponentially with the number of relations joined, and therefore do not scale to large datasets. In this section we describe a "virtual join" algorithm that computes the required data tables without the quadratic cost of materializing a cross-product.

Our first observation is that Bayes net learners do not require the entire extended table, but only the *sufficient statistics*, namely the counts of how many times each combination of values occurs in the data [11]. For instance, the attribute-relationship table of Figure 1, each combination is observed exactly once. Our second observation is that to compute the sufficient statistics, it suffices to compute the *frequencies* of value combinations rather than the counts, because counts can be obtained from frequencies by multiplying with the appropriate domain sizes. An efficient virtual join algorithm for computing frequencies in relational data was recently published by Schulte *et al.* [18]. Their algorithm is based on the fast Möbius transform (FMT). We outline it briefly.

Consider a set of relationship indicator nodes $R_1 = \cdot, R_2 = \cdot, \ldots, R_m$ and attribute nodes $f_1, \ldots, f_j$. The sufficient statistics correspond to a joint distribution over these random variables.

$$P(R_1 = \cdot, R_2 = \cdot, \ldots, R_m = \cdot; f_1 = \cdot, \ldots, f_j = \cdot).$$

Our goal is to compute sufficient statistics of this form for relational data. The FMT allows us to efficiently find sufficient statistics for binary random variables. We apply it with a fixed set of values for the attribute nodes, which corresponds to a joint distribution over the m Boolean relationship random variables:

$$P(R_1 = \cdot, R_2 = \cdot, \ldots, R_m = \cdot; f_1 = v_1, \ldots, f_j = v_j).$$

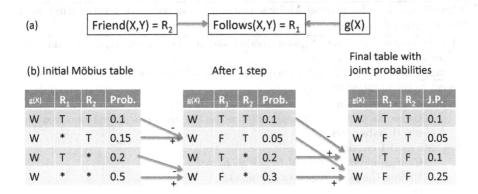

Fig. 4. (a) A Bayes net with two relationship nodes. (b) An illustrative trace of the fast Möbius transform.

The FMT uses the local update operation

$$P(R = F, \mathbf{R}) = P(\mathbf{R}) - P(R = T, \mathbf{R}) \tag{1}$$

where $\mathbf{R}$ is a conjunction of relationship specifications, possibly with both positive and negative relationships. The equation continues to hold if we extend relationship specifications with any fixed set of value assignments $f_1 = v_1, \ldots, f_j = v_j$ to attribute functor nodes $f_1, \ldots, f_j$. Using the convention that $R = *$ means that the value of relationship R is unspecified, the equation (1) can be rewritten as

$$P(R = F, \mathbf{R}) = P(R = *, \mathbf{R}) - P(R = T, \mathbf{R}). \tag{2}$$

The FMT begins with an initial table of sufficient statistics where all relationship nodes have the value T or $*$ but not F. Since these sufficient statistics do not involve false relationships, they can be computed efficiently from a relational database using table joins. The procedure then goes through the relationship nodes $R_1, \ldots, R_m$ in order, at stage i replacing all occurrences of $R_i = *$ with $R_i = F$, and applying the local update equation to obtain the probability value for the modified row. At termination, all $*$ values have been replaced by F and the table specifies all joint frequencies as required. Algorithm 2 gives pseudo code and Figure 4 presents an example of the transform step. For example, the probability entry for the second row of the middle table is computed by applying the equation

$$P(R_1 = F, R_2 = T; g(X) = W) =$$
$$P(R_1 = *, R_2 = T; g(X) = W) - P(R_1 = T, R_2 = T; g(X) = W) =$$
$$0.15 - 0.1 = 0.05$$

which is an instance of Equation (2).

Algorithm 2. The inverse Möbius transform for parameter estimation in a Parametrized Bayes Net

Input: database $\mathcal{D}$; a set of nodes divided into attribute nodes $f_1, \ldots, f_j$ and relationship nodes $R_1, \ldots, R_m$.
Output: joint probability table specifying the data frequencies for each joint assignment to the input nodes.

1: **for all** attribute value assignments $f_1 := v_1, \ldots, f_j := v_j$ **do**
2:    initialize the table: set all relationship nodes to either T or $*$; find joint frequencies with data queries.
3:    **for** $i = 1$ to m **do**
4:        Change all occurrences of $R_i = *$ to $R_i = F$.
5:        Update the joint frequencies using (1).
6:    **end for**
7: **end for**

7   Conclusion

We described different methods for extending relational Bayes net learning to correlations involving links. Statistical measures indicate that Bayes net methods succeed in finding relevant correlations. There is a trade-off between statistical power and computational feasibility (full table search vs constrained search). Hierarchical search often does well on both dimensions, but needs to be extended with a pruning step to eliminate redundant edges.

A key issue for scalability is that most of the learning time is taken up by forming table joins, whose size is the cross product of entity tables. These table joins provide the sufficient statistics required in model selection. To improve scalability, computing sufficient statistics needs to be feasible for cross product sizes in the millions or more. A solution is applying virtual join methods that compute sufficient statistics without materializing table joins, such as the Fast Möbius Transform [17,22].

A valuable direction for future work is to compare learning link correlations with directed and undirected models, such as Markov Logic Networks [2]. As we explained in Section 2, current relational learners for undirected models do not scale to most of our datasets. One option is to subsample the datasets so that we can compare the statistical power of directed and undirect learning methods independently of scalability issues. Khosravi *et al.* were able to obtain structure learning results for Alchemy [7], but did not evaluate the models with respect to link correlations. For the MLN-Boost system, we were able to obtain preliminary results on several benchmark databases (including Mutagenesis and Hepatitis), by selecting the right subset of target predicates. MLN-Boost is the current state-of-the-art learner for Markov Logic Networks [8]. The Bayes net models were competitive with the MLN-Boost models on a standard cross-validation measure of predictive accuracy.

References

1. Chickering, D.: Optimal structure identification with greedy search. Journal of Machine Learning Research 3, 507–554 (2003)
2. Domingos, P., Lowd, D.: Markov Logic: An Interface Layer for Artificial Intelligence. Morgan and Claypool Publishers (2009)
3. Getoor, L., Friedman, N., Koller, D., Pfeffer, A., Taskar, B.: Probabilistic relational models. In: Introduction to Statistical Relational Learning, ch. 5, pp. 129–173. MIT Press (2007)
4. Getoor, L., Taskar, B., Koller, D.: Selectivity estimation using probabilistic models. ACM SIGMOD Record 30(2), 461–472 (2001)
5. Heckerman, D., Geiger, D., Chickering, D.: Learning Bayesian networks: The combination of knowledge and statistical data. Machine Learning 20, 197–243 (1995)
6. Khosravi, H., Man, T., Hu, J., Gao, E., Schulte, O.: Learn and join algorithm code, http://www.cs.sfu.ca/~oschulte/jbn/
7. Khosravi, H., Schulte, O., Man, T., Xu, X., Bina, B.: Structure learning for Markov logic networks with many descriptive attributes. In: AAAI, pp. 487–493 (2010)
8. Khot, T., Natarajan, S., Kersting, K., Shavlik, J.W.: Learning Markov logic networks via functional gradient boosting. In: ICDM, pp. 320–329. IEEE Computer Society (2011)
9. Khot, T., Shavlik, J., Natarajan, S.: Boostr (2013), http://pages.cs.wisc.edu/~tushar/Boostr/
10. Natarajan, S., Khot, T., Kersting, K., Gutmann, B., Shavlik, J.W.: Gradient-based boosting for statistical relational learning: The relational dependency network case. Machine Learning 86(1), 25–56 (2012)
11. Neapolitan, R.E.: Learning Bayesian Networks. Pearson Education (2004)
12. Poole, D.: First-order probabilistic inference. In: IJCAI, pp. 985–991 (2003)
13. Russell, S., Norvig, P.: Artificial Intelligence: A Modern Approach. Prentice Hall (2010)
14. Schulte, O.: A tractable pseudo-likelihood function for Bayes nets applied to relational data. In: SIAM SDM, pp. 462–473 (2011)
15. Schulte, O.: Challenge paper: Marginal probabilities for instances and classes. In: ICML-SRL Workshop on Statistical Relational Learning (June 2012)
16. Schulte, O., Khosravi, H.: Learning graphical models for relational data via lattice search. Machine Learning 88(3), 331–368 (2012)
17. Schulte, O., Khosravi, H., Kirkpatrick, A., Gao, T., Zhu, Y.: Modelling relational statistics with bayes nets. In: Inductive Logic Programming, ILP (2012)
18. Schulte, O., Khosravi, H., Kirkpatrick, A., Gao, T., Zhu, Y.: Modelling relational statistics with bayes nets. Machine Learning (2013) (forthcoming)
19. Sun, Y., Han, J.: Mining Heterogeneous Information Networks: Principles and Methodologies, vol. 3. Morgan & Claypool Publishers (2012)
20. The Tetrad Group. The Tetrad project (2008), http://www.phil.cmu.edu/projects/tetrad/
21. Ullman, J.D.: Principles of Database Systems, 2nd edn. W. H. Freeman & Co. (1982)
22. Yin, X., Han, J., Yang, J., Yu, P.S.: Crossmine: Efficient classification across multiple database relations. In: ICDE, pp. 399–410. IEEE Computer Society (2004)

Concurrent Reasoning with Inference Graphs

Daniel R. Schlegel and Stuart C. Shapiro

Department of Computer Science and Engineering
University at Buffalo, Buffalo NY 14260, USA
{drschleg,shapiro}@buffalo.edu

Abstract. Since their popularity began to rise in the mid-2000s there has been significant growth in the number of multi-core and multi-processor computers available. Knowledge representation systems using logical inference have been slow to embrace this new technology. We present the concept of inference graphs, a natural deduction inference system which scales well on multi-core and multi-processor machines. Inference graphs enhance propositional graphs by treating propositional nodes as tasks which can be scheduled to operate upon messages sent between nodes via the arcs that already exist as part of the propositional graph representation. The use of scheduling heuristics within a prioritized message passing architecture allows inference graphs to perform very well in forward, backward, bi-directional, and focused reasoning. Tests demonstrate the usefulness of our scheduling heuristics, and show significant speedup in both best case and worst case inference scenarios as the number of processors increases.

1 Introduction

Since at least the early 1980s there has been an effort to parallelize algorithms for logical reasoning. Prior to the rise of the multi-core desktop computer, this meant massively parallel algorithms such as that of [3] on the (now defunct) Thinking Machines Corporation's Connection Machine, or using specialized parallel hardware which could be added to an otherwise serial machine, as in [10]. Parallel logic programming systems designed during that same period were less attached to a particular parallel architecture, but parallelizing Prolog (the usual goal) is a very complex problem [17], largely because there is no persistent underlying representation of the relationships between predicates. Parallel Datalog has been more successful (and has seen a recent resurgence in popularity [7]), but is a much less expressive subset of Prolog. Recent work in parallel inference using statistical techniques has returned to large scale parallelism using GPUs, but while GPUs are good at statistical calculations, they do not do logical inference well [24].

We present *inference graphs* [16], a graph-based natural deduction inference system which lives within a KR system, and is capable of taking advantage of multiple cores and/or processors using concurrent processing techniques rather than parallelism [21]. We chose to use natural deduction inference, despite the existence of very well performing refutation based theorem provers, because our system is designed to be able to perform forward inference, bi-directional inference [19], and focused reasoning in addition to the backward inference used in resolution. Natural deduction also allows

M. Croitoru et al. (Eds.): GKR 2013, LNAI 8323, pp. 138–164, 2014.

formulas generated during inference to be retained in the KB for later re-use, whereas refutation techniques always start by assuming the negation of the formula to be derived, making intermediate derivations useless for later reasoning tasks. In addition, our system is designed to allow formulas to be disbelieved, and to propagate that disbelief to dependent formulas. We believe inference graphs are the only concurrent inference system with all these capabilities.

Inference graphs are, we believe, unique among logical inference systems in that the graph representation of the KB is the same structure used for inference. Because of this, the inference system needn't worry about maintaining synchronicity between the data contained in the inference graph, and the data within the KB. This contrasts with systems such as [23] which allow queries to be performed upon a graph, not within it. More similar to our approach is that of Truth Maintenance Systems [4] (TMSes), which provide a representation of knowledge based on justifications for truthfulness. The TMS infers within itself the truth status of nodes based on justifications, but the justifications themselves must be provided by an external inference engine, and the truth status must then be reported back to the inference engine upon request.

Drawing influences from multiple types of TMS [4,8,12], RETE networks [5], and Active Connection Graphs [13], and implemented in Clojure [6], inference graphs are an extension of propositional graphs allowing messages about assertional status and inference control to flow through the graph. The existing arcs within propositional graphs are enhanced to carry messages from antecedents to rule nodes, and from rule nodes to consequents. A rule node in an inference graph combines messages and implements introduction and elimination rules to determine if itself or its consequents are true or negated.

In Sect. 2 we review propositional graphs and introduce an initial example, followed by our introduction to inference graphs in Sect. 3. Section 4 explains how the inference graphs are implemented in a concurrent processing system. In Sect. 5 we present a second illustrative example. We evaluate the implementation in Sect. 6 and finally conclude with Sect. 7.

2 Propositional Graphs

Propositional graphs in the tradition of the SNePS family [20] are graphs in which every well-formed expression in the knowledge base, including individual constants, functional terms, atomic formulas, and non-atomic formulas (which we will refer to as "rules"), is represented by a node in the graph. A rule is represented in the graph as a node for the rule itself (henceforth, a *rule node*), nodes for the argument formulas, and arcs emanating from the rule node, terminating at the argument nodes. Arcs are labeled with an indication of the role the argument plays in the rule, itself. Every node is labeled with an identifier. Nodes representing individual constants, proposition symbols, function symbols, or relation symbols are labeled with the symbol itself. Nodes representing functional terms or non-atomic formulas are labeled $wft\,i$, for some integer, i.[1] An exclamation mark, "!", is appended to the label if the proposition is true in the KB. No two nodes represent syntactically identical expressions; rather, if there are

---

[1] "wft" rather than "wff" for reasons that needn't concern us in this paper.

multiple occurrences of one subexpression in one or more other expressions, the same node is used in all cases.

In this paper, we will limit our discussion to inference over formulas of logic which do not include variables or quantifiers.[2] Specifically, we have implemented introduction and elimination rules for the set-oriented connectives andor, and thresh [18], and the elimination rules for numerical entailment. The andor connective, written (andor $(i\ j)\ p_1 \ldots p_n$), $0 \le i \le j \le n$, is true when at least i and at most j of $p_1 \ldots p_n$ are true (that is, an andor may be introduced when those conditions are met). It generalizes and $(i = j = n)$, or $(i = 1, j = n)$, nand $(i = 0, j = n-1)$, nor $(i = j = 0, n > 1)$, xor $(i = j = 1)$, and not $(i = j = 0, n = 1)$. For the purposes of andor-elimination, each of $p_1 \ldots p_n$ may be treated as an antecedent or a consequent, since when any j formulas in $p_1 \ldots p_n$ are known to be true (the antecedents), the remaining formulas (the consequents) can be inferred to be negated, and when any $n - i$ arguments are known to be false, the remaining arguments can be inferred to be true. For example with xor, a single true formula causes the rest to become negated, and if all but one are found to be negated, the remaining one can be inferred to be true. The thresh connective, the negation of andor, and written (thresh $(i\ j)\ p_1 \ldots p_n$), $0 \le i \le j \le n$, is true when either fewer than i or more than j of $p_1 \ldots p_n$ are true. The thresh connective is mainly used for equivalence (iff), when $i = 1$ and $j = n - 1$. As with andor, for the purposes of thresh-elimination, each of $p_1 \ldots p_n$ may be treated as an antecedent or a consequent. Numerical entailment is a generalized entailment connective, written (=> i (setof $a_1 \ldots a_n$) (setof $c_1 \ldots c_m$)) meaning if at least i of the antecedents, $a_1 \ldots a_n$, are true then all of the consequents, $c_1 \ldots c_m$, are true. The initial example and evaluations in this paper will make exclusive use of two special cases of numerical entailment – or-entailment, where $i = 1$, and and-entailment, where $i = n$ [20].

In our initial example we will use two rules, one using an and-entailment and one using an or-entailment: (if (setof a b c) d), meaning that whenever a, b, and c are true then d is true; and (v=> (setof d e) f), meaning that whenever d or e is true then f is true. In Fig. 1 wft1 represents the and-entailment (if (setof a b c) d). The identifier of the rule node wft1 is followed by an exclamation point, "!", to indicate that the rule (well-formed formula) is asserted – taken to be true. The antecedents, a, b, and c are connected to wft1 by arcs labeled with $\land$ant, indicating they are antecedents of an and-entailment. An arc labeled cq points from wft1 to d, indicating d is the consequent of the rule. wft2 represents the or-entailment (v=> (setof d e) f). It is assembled in a similar way to wft1, but the antecedents are labeled with $\lor$ant, indicating this rule is an or-entailment. While much more complex examples are possible (and one is shown in Section 5), we will use this rather simplistic one in our initial example since it illustrates the concepts presented in this paper without burdening the reader with the details of the, perhaps less familiar, andor and thresh connectives.

The propositional graphs are fully indexed, meaning that not only can the node at the end of an arc be accessed from the node at the beginning of the arc, but the node at the beginning of the arc can be accessed from the node at the end of the arc. For example, it is possible to find all the rule nodes in which d is a consequent by following

---

[2] The system has been designed to be extended to a logic with quantified variables in the future.

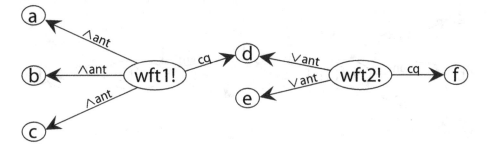

Fig. 1. Propositional graph for the assertions that if a, b, and c are true, then d is true, and if d or e are true, then f is true

cq arcs backward from node d (We will refer to this as following the reverse cq arc.), and it is possible to find all the or-entailment rule nodes in which d is an antecedent by following reverse ∨ant arcs from node d.

3 Inference Graphs

Inference graphs are an extension of propositional graphs to allow deductive reasoning to be performed in a concurrent processing system. Unlike many KR systems which have different structures for representation and inference, inference graphs serve as both the representation and the inference mechanism.

To create an inference graph, certain arcs and certain reverse arcs in the propositional graph are augmented with *channels* through which information can flow. Channels come in two forms. The first type, *i-channels*, are added to the reverse antecedent arcs. These are called i-channels since they carry messages reporting that "I am true" or "I am negated" from the antecedent node to the rule node. Channels are also added to the consequent arcs, called *u-channels*,[3] since they carry messages to the consequents which report that "you are true" or "you are negated." Rules are connected by shared subexpressions, as wft1 and wft2 are connected by the node d.

Figure 2 shows the propositional graph of Fig. 1 with the appropriate channels illustrated. i-channels (dashed arcs) are drawn from a, b, and c to wft1, and from d, and e to wft2. These channels allow the antecedents to report to a rule node when it (or its negation) has been derived. u-channels (dotted arcs) are drawn from wft1 to d and from wft2 to f so that rule nodes can report to consequents that they have been derived.

Each channel contains a valve. Valves enable or prevent the flow of messages forward through the graph's channels. When a valve is closed, any new messages which arrive at it are added to a waiting set. When a valve opens, messages waiting behind it are sent through. Often we will speak of a channel being open or closed, where that is intended to mean that the valve in the channel is open or closed.

---

[3] u-channels and u-infer messages were previously called y-channels and Y-INFER messages in [14,16].

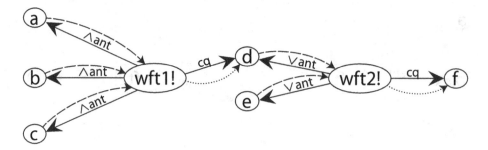

Fig. 2. The same propositional graph from Fig. 1 with the appropriate channels added. Channels represented by dashed lines are i-channels and are drawn from antecedents to rule nodes. Channels represented by dotted lines are u-channels and are drawn from rule nodes to consequents.

Inference graphs are capable of forward, backward, bi-directional, and focused inference. In forward inference, messages flow forward through the graph, ignoring any valves they reach. In backward inference, messages are sent backward through incoming channels to open valves, and hence other messages about assertions are allowed to flow forward only along the appropriate paths. Normally when backward inference completes, the valves are closed. Focused inference can be accomplished by allowing those valves to remain open, thus allowing messages reporting new assertions to flow through them to answer previous queries. Bi-directional inference can begin backward inference tasks to determine if an unasserted rule node reached by a forward inference message can be derived.

3.1 Messages

Messages of several types are transmitted through the inference graph's channels, serving two purposes: relaying newly derived information, and controlling the inference process. A message can be used to relay the information that its origin has been asserted or negated (an i-infer message), that its destination should now be asserted or negated (u-infer), or that its origin has either just become unasserted or is no longer sufficiently supported (unassert). These messages all flow forward through the graph. Other messages flow backward through the graph controlling inference by affecting the channels: backward-infer messages open them, and cancel-infer messages close them. Messages are defined as follows:

$$< origin, support, type, assertStatus?, fwdInfer?, priority >$$

where *origin* is the node which produced the message; *support* is the set of support which allowed the derivation producing the message (for ATMS-style belief revision [8,11]); *type* is the type of the message (such as i-infer or backward-infer, described in detail in the next several subsections); *fwdInfer?* states whether this message is part of a forward-inference task; *assertStatus?* is whether the message is reporting about a newly true or newly false node; and *priority* is used in scheduling the consumption of messages (discussed further in Sect. 4). We define the different types of messages below, while the actions the receipt of those messages cause are discussed in Secs. 3.2 and 3.3.

i-infer. When a node is found to be true or false, an i-infer message reflecting this new assertional status is submitted to its outgoing i-channels. These messages are sent from antecedents to rule nodes. An i-infer message contains a support set which consists of every node used in deriving the originator of the message. These messages optionally can be flagged as part of a forward inference operation, in which case they treat any closed valves they reach as if they were open. The priority of an i-infer message is one more than that of the message that caused the change in assertional status of the originator. Section 4 will discuss why this is important.

u-infer. Rule nodes which have just learned enough about their antecedents to fire send u-infer messages to each of their consequents, informing them of what their new assertional status is – either true or false.[4] As with i-infer messages, u-infer messages contain a support set, can be flagged as being part of a forward inference operation, and have a priority one greater than the message that preceded it.

backward-infer. When it is necessary for the system to determine whether a node can be derived, backward-infer messages are used to open channels to rules that node may be the consequent of, or, if the node is a rule node, to open channels of antecedents to derive the rule. These messages set up a backward inference operation by passing backward through channels and opening any valves they reach. In a backward inference operation, these messages are generally sent recursively backward through the network until valves are reached which have waiting i-infer or u-infer messages. The priority of these messages is lower than any inference tasks which may take place. This allows any messages waiting at the valves to flow forward immediately and begin inferring new formulas towards the goal of the backward inference operation.

cancel-infer. Inference can be canceled either in whole by the user or in part by a rule which determines that it no longer needs to know the assertional status of some of its antecedents.[5] cancel-infer messages are sent from some node backward through the graph. These messages are generally sent recursively backward through the network to prevent unnecessary inference. These messages cause valves to close as they pass through, and cancel any backward-infer messages scheduled to re-open those same valves, halting inference. cancel-infer messages always have a priority higher than any inference task.

unassert. Each formula which has been derived in the graph has one or more sets of support. For a formula to remain asserted, at least one of its support sets must have all of

---

[4] For example a rule representing the exclusive or of a and b could tell b that it is true when it learns that a is false, or could tell b that it is false when it learns that a is true.

[5] We recognize that this can, in some cases, prevent the system from automatically deriving a contradiction.

its formulas asserted as well. When a formula is unasserted by a human (or, eventually, by belief revision), a message must be sent forward through the graph to recursively unassert any formulas which have the unasserted formula in each of their support sets, or is the consequent of a rule which no longer has the appropriate number of positive or negative antecedents. `unassert` messages have top priority, effectively pausing all other inference, to help protect the consistency of the knowledge base.

3.2 Rule Node Inference

Inference operations take place in the rule nodes. When an `i-infer` message arrives at a rule node the message is translated into *Rule Use Information*, or RUI [2]. A RUI is defined as:

$$< pos, neg, flaggedNS, support >$$

where *pos* and *neg* are the number of known true ("positive") and negated ("negative") antecedents of the rule, respectively; the *flaggedNS* is the *flagged node set*, which contains a mapping from each antecedent with a known truth value to its truth value, and *support* is the set of support, the set of formulas which must be true for the assertional statuses in the *flaggedNS* to hold.

All RUIs created at a node are cached. When a new one is made, it is combined with any already existing ones – *pos* and *neg* from the RUIs are added and the set of support and flagged node set are combined. The output of the combination process is a set of new RUIs created since the message arrived at the node. The *pos* and *neg* portions of the RUIs in this set are used to determine if the rule node's inference rules can fire. A disadvantage of this approach is that some rules are difficult, but not impossible, to implement, such as negation introduction and proof by cases. For us, the advantages in capability outweigh the difficulties of implementation. If the RUI created from a message already exists in the cache, no work is done. This prevents re-derivations, and can cut cycles. When a rule fires, new `u-infer` messages are sent to the rule's consequents, informing them whether they should be asserted or negated.

Figure 3 shows the process of deriving f in our initial example. We assume backward inference has been initiated, opening all the valves in the graph. First, in Fig. 3a, messages about the truth of a, b, and c flow through i-channels to `wft1`. Since `wft1` is and-entailment, each of its antecedents must be true for it to fire. Since they are, in Fig. 3b the message that d is true flows through `wft1`'s u-channel. d becomes asserted and reports its new status through its i-channel (Fig. 3c). In Fig. 3d, `wft2` receives this information, and since it is an or-entailment rule and requires only a single antecedent to be true for it to fire, it reports to its consequents that they are now true, and cancels inference in e. Finally, in Fig. 3e, f is asserted, and inference is complete.

3.3 Inference Segments

The inference graph is divided into *inference segments* (henceforth, *segments*). A segment represents the operation – from receipt of a message to sending new ones – which occurs in a node. Valves delimit segments, as seen in Fig. 4. The operation of a node

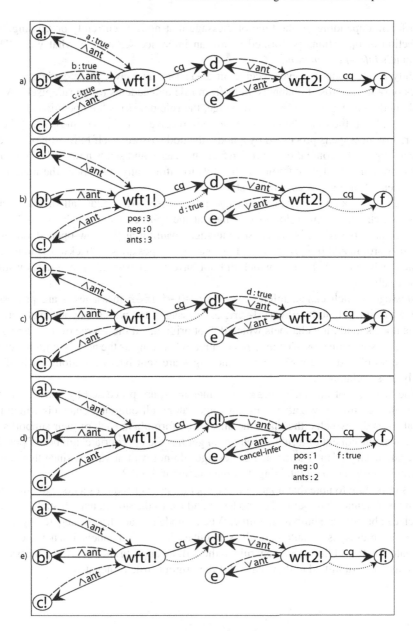

Fig. 3. a) Messages are passed from a, b, and c to wft1. b) wft1 combines the messages from a, b, and c to find that it has 3 positive antecedents, of a total of 3. The and-entailment can fire, so it sends a message through its u-channel informing its consequent, d, that it has been derived. c) d receives the message that it is asserted and sends messages through its i-channel. d) wft2 receives the message that d is asserted. Only one true antecedent is necessary for or-entailment elimination, so it sends a message through its u-channels that its consequent, f, is now derived. It also cancels any inference in its other antecedents by sending a cancel-infer message to e. e) f is derived.

is different depending on the type of message that node is currently processing. The collection of operations performed within an inference segment is what we call the segment's *inference function*.

Only a rule node can ever receive an i-infer message, since i-infer messages only flow across i-channels, built from rule antecedents to the rule itself. When a rule node receives an i-infer message, the rule performs the operation discussed in Sec. 3.2. On the other hand, any node may receive a u-infer message. When a u-infer message is processed by a node, the node asserts itself to be true or negated (depending on the content of the u-infer message), and sends new i-infer messages through any of its outgoing i-channels to inform any rule nodes the node is an antecedent of of it's new assertional status.

When a backward-infer message is processed by a node, all of that node's incoming channels which do not originate at the source of the backward-infer message are opened. This act causes messages waiting at the now-opened valves to flow forward. In addition, backward-infer messages are sent backward along each of the newly-opened channels which did not have i- or u-infer messages waiting at their valves.

Messages which cancel inference – cancel-infer messages – are processed somewhat similarly to backward-infer messages, but with the opposite effect. All of the node's incoming channels which do not originate at the source of the cancel-infer message are closed (if they're not already), as long as the node's outgoing channels are all closed. cancel-infer messages are sent backward along each of the newly-closed channels.

The last type of message, unassert messages, are processed by unasserting the node, and sending new unassert messages along all outgoing channels which terminate at nodes which either have the unasserted formula in each of their support sets, or are the consequent of a rule which no longer has the appropriate number of positive or negative antecedents. unassert messages do not wait at valves, since they are of critical importance to maintaining the consistency of the KB.

It's important to note that none of these operations ever requires a node to build new channels. Channels are built when nodes are added to the graph, and all such channels which can be built are built at that time. When a node is made true or negated by a user, i-infer messages are automatically sent out each of its i-channels. If a new i-channel is created by the addition of a new rule to the KB, an appropriate i-infer message is automatically submitted to that channel if the origin is true or negated.

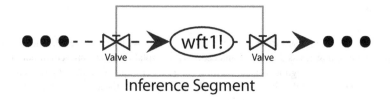

Fig. 4. A single inference segment is shown in the gray bounding box

4   Concurrent Reasoning

The inference graph's structure lends itself naturally to concurrent inference. It is clear in our initial example that if inference were required to derive a, b, and c that each of those inference tasks could be running concurrently. After each of a, b, and c were asserted, messages would be sent to wft1, as in our example. The RUIs generated from the messages would then need to be combined. Since there is shared state (the RUI cache) we perform the combination of RUIs synchronously using Clojure's Software Transactional Memory, guaranteeing we don't "lose" results. We need not concern ourselves with the actual order in which the RUIs are combined, since the operation is commutative, meaning we don't need to maintain a queue of changes to the RUI cache.

In addition to the RUI cache, we must also update the set of asserted formulas whenever a formula is newly derived. We use the same technique as above for this, again recognizing that the order in which formulas are asserted is not important. It is possible to perform these assertions in a thread outside of the inference task, but tests show no advantage in doing so,[6] especially in the case where we have as many (or more) inference threads as CPUs.

In order to perform inference concurrently, we aim to execute the inference functions of multiple segments at the same time, and do so in an efficient manner through the use of scheduling heuristics. A *task* is the application of a segment's inference function to a message. Tasks are created whenever a message passes the boundary between segments. For i-infer and u-infer messages, that's when they pass through a valve, or for unassert messages when they ignore a valve. For the backward-infer and cancel-infer messages it's when they cross a valve backward. When tasks are created they enter a global prioritized queue, where the priority of the task is the priority of the message. The task which causes a message to cross between segments is responsible for creating the new task for that message and adding it to the queue. When the task is executed, the appropriate operation is performed as described above, and any newly generated messages allow the process to repeat.

4.1   Scheduling Heuristics

The goal of any inference system is to infer the knowledge requested by the user. If we arrange an inference graph so that a user's request (in backward inference) is on the right, and channels flow from left to right wherever possible (the graph may contain cycles), we can see this goal as trying to get messages from the left side of the graph to the right side of the graph. We, of course, want to do this as quickly as possible.

Every inference operation begins processing messages some number of levels to the left of the query node. Since there are a limited number of tasks which can be running

---

[6] This can actually reduce performance by causing introduction rules to fire more than once. Consider two i-infer messages arriving at an unasserted rule node nearly simultaneously. Assume the first message triggers the rule's introduction. The rule requests, asynchronously, to be asserted, then sends appropriate messages through the its i- and u-channels. The second message arrives, but the node's assertion request has not completed yet, causing duplication of work. This can happen (rarely) in the current concurrency model, but is far more likely when it's uncertain when the actual assertion will occur.

at once due to hardware limitations, we must prioritize their execution, remove tasks which we know are no longer necessary, and prevent the creation of unnecessary tasks. Therefore,

1. tasks for relaying newly derived information using segments to the right are executed before those to the left,
2. once a node is known to be true or false, all tasks still attempting to derive it are canceled, as long as their results are not needed elsewhere, and all channels pointing to it which may still derive it are closed.
3. once a rule fires, all tasks for potential antecedents of that rule still attempting to satisfy it are canceled, as long as their results are not needed elsewhere, and all channels from antecedents which may still satisfy it are closed.

Together, these three heuristics ensure that messages reach the query as quickly as possible, and time is not wasted deriving unnecessary formulas. The priorities of the messages (and hence, tasks) allow us to reach these goals. All unassert messages have the highest priority, followed by all cancel-infer messages. Then come i-infer and u-infer messages. backward-infer messages have the lowest priority. As i-infer and u-infer messages flow to the right, they get higher priority, but their priorities remain lower than that of cancel-infer messages. In forward inference, i-infer and u-infer messages to the right in the graph always have higher priority than those to the left, since the messages all begin flowing from a common point. In backward inference, the priorities of backward-infer, and i-infer, and u-infer messages work together to derive a query formula as quickly as possible: since backward-infer messages are of the lowest priority, those i-infer and u-infer messages waiting at valves which are nearest to the query formula begin flowing forward before valves further away are opened. This, combined with the increasing priority of i-infer and u-infer messages ensure efficient derivation. In short, the closest possible path to the query formula is always attempted first in backward inference.

The usefulness of cancel-infer can be seen if you consider what would happen in the example in Fig. 3 if e were also asserted. Remember that the cancel-infer messages close valves in channels they passes through, and are passed backward further when a node has received the same number of cancellation messages as it has outgoing channels. In this example, backward inference messages would reach wft2, then d and e. The message that e is asserted would flow forward through e's i-channel to wft2, which would in turn both send a message resulting in the assertion of f, and cancel inference going on further left in the graph, cutting off wft1's inference since it is now unnecessary.

The design of the system therefore ensures that the tasks executing at any time are the ones closest to deriving the goal, and tasks which will not result in useful information towards deriving the goal are cleaned up. Additionally, since nodes "push" messages forward through the graph instead of "pulling" from other nodes, it is not possible to have tasks running waiting for the results of other rule nodes' tasks. Thus, deadlocks are impossible, and bottlenecks can only occur when multiple threads are making additions to shared state simultaneously.

5   An Illustrative Example

The initial example we have discussed thus far was useful for gaining an initial understanding of the inference graphs, but it is not sophisticated enough to show several interesting properties of the graphs. The following example shows how the inference graph can be used for rule introduction, deriving negations, the use of ground predicates, and the set-oriented logical connectives such as andor, xor, and iff discussed in Sect. 2.

Inspired by L. Frank Baum's *The Wonderful Wizard of Oz* [1], we consider a scene in which Dorothy and her friends are being chased by Kalidas – monsters with the head of a tiger and the body of a bear. In the world of Oz, whenever Toto becomes scared, Dorothy carries him, and that's the only time she carries him. If Toto walks, he is not carried, and if he is carried, he does not walk. Toto becomes scared when Dorothy is being chased. Since Dorothy has only two hands, she is capable of either carrying the Scarecrow (who is large, and requires both hands), or between 1 and 2 of the following items: Toto, her full basket, and the Tin Woodman's oil can. In our example, the Tin Woodman is carrying his own oil can. Only one of Dorothy, the Scarecrow, or the Tin Woodman can carry the oil can.

The relevant parts of this scene are represented below in their logical forms.

```
;;; Dorothy can either carry the scarecrow,
;;;   or carry one or two objects from the list:
;;;     her full basket, Toto, oil can.
(xor (Carries Dorothy Scarecrow)
     (andor (1 2) (Carries Dorothy FullBasket)
                  (Carries Dorothy Toto)
                  (Carries Dorothy OilCan)))

;;; Either Dorothy, the Tin Woodman, or the Scarecrow
;;;     carry the Oil Can.
(xor (Carries Dorothy OilCan)
     (Carries TinWoodman OilCan)
     (Carries Scarecrow OilCan))

;;; Either Dorothy carries Toto, or Toto walks.
(xor (Carries Dorothy Toto) (Walks Toto))

;;; Dorothy carries Toto if and only if Toto is scared.
(iff (Scare Toto) (Carries Dorothy Toto))

;;; Toto gets scared if Dorothy is being chased.
(if (Chase Dorothy) (Scare Toto))

;;; The Tin Woodman is carrying his Oil Can.
(Carries TinWoodman OilCan)
```

```
;;; Dorothy is being chased.
(Chase Dorothy)
```

We can then wonder, "Is Dorothy carrying the Scarecrow?" According to the rules of these connectives, as discussed in Section 2, we should be able to derive that this is not the case. This can be derived by hand as follows:

```
1) Since (if (Chase Dorothy) (Scare Toto))
      and (Chase Dorothy),
   infer (Scare Toto)
                           by Implication Elimination.
2) Since (iff (Carries Dorothy Toto) (Scare Toto))
      and (Scare Toto)
   infer (Carries Dorothy Toto)
                           by Equivalence Elimination.
3) Since (xor (Carries Scarecrow OilCan)
              (Carries Dorothy OilCan)
              (Carries TinWoodman OilCan))
      and (Carries TinWoodman OilCan)
   infer (not (Carries Dorothy OilCan))
                           by Xor Elimination.
4) Since (Carries Dorothy Toto)
      and (not (Carries Dorothy OilCan))
   infer (andor (1 2) (Carries Dorothy FullBasket)
                      (Carries Dorothy OilCan)
                      (Carries Dorothy Toto))
                           by Andor Introduction.
5) Since (xor
              (andor (1 2) (Carries Dorothy FullBasket)
                           (Carries Dorothy OilCan)
                           (Carries Dorothy Toto))
              (Carries Dorothy Scarecrow))
      and (andor (1 2) (Carries Dorothy FullBasket)
                       (Carries Dorothy OilCan)
                       (Carries Dorothy Toto))
   infer (not (Carries Dorothy Scarecrow))
                           by Xor Elimination.
```

The inference graphs are able to reach the same conclusion by applying these same rules of inference. The inference graph for the above example is displayed in Fig. 5. Since this example makes use of ground predicate logic (instead of only proposition symbols, as used in the initial example), we have had to define the arc labels used to identify the arguments of those formulas in the graph.[7] The `Carries` relation has two arguments, `carrier` and `carried`, where the `carrier` is the entity carrying the `carried` object. The `Walks` relation has only the `selfMover` argument – the entity doing the walking. Both the `Chase` and `Scare` relations have only one argument. The `Chase` relation has a `theme` – the one being chased – and the `Scare` relation has an `experiencer` – the one who becomes scared. In order to make it clear in the graph what the act the `theme` or `experiencer` is involved in, we add an additional arc labeled `act` pointing to a node whose symbol is the relation name.[8]

In Fig. 5, `wft6` represents the proposition that either Dorothy carries the Scarecrow (`wft1`), or `wft5`, that Dorothy carries between 1 and 2 of the items: the oil can (`wft2`), her full basket (`wft3`), and Toto (`wft4`). `wft2`, `wft8` and ,`wft9` respectively represent the propositions that Dorothy, the Scarecrow, and the Tin Woodman carry the oil can, and `wft10` is the exclusive disjunction of those formulas. The exclusive-or rule node `wft14` represents the proposition that either Toto walks (`wft13`) or Toto is carried by Dorothy (`wft4`). `wft12` expresses that Dorothy carries Toto (`wft4`) if and only if Toto is scared (`wft11`). Finally, the relation expressing that if Dorothy is being chased (`wft15`), then Toto is scared (`wft13`) is represented by `wft16`. Notice that, as indicated by the appended "!", `wft6`, `wft9`, `wft10`, `wft12`, `wft14`, `wft15`, and `wft16`, are asserted, but none of the other wfts are.

Channels have been drawn on the graph as described earlier. Since `andor` and `thresh` do not have pre-defined antecedents and consequents (as the various types of entailment do), each of the arguments in one of these rules must, in the inference graph, have an i-channel drawn from the argument to the rule, and a u-channel drawn from the rule to the argument. This way each argument can inform the rule when it has a new assertional status, and the rule can inform each argument about a new assertional status it should adopt based on the rule firing.

For the purposes of this example, we will make two assumptions: first that there are two processors being used (*a* and *b*), and second that any two tasks which begin on the two CPUs simultaneously, end at the same time as well. Figure 6 shows the first set of processing steps used in the derivation. Processing steps in this figure are labeled one through five, with "a" and "b" appended to the label where necessary to denote the CPU in use for ease of reference to the diagram. The step labels are placed at the nodes, since a task stretches from valve-to-valve, encompassing a single node. We'll discuss the inference process as if the nodes themselves are added to the task queue for easier reading, when what we really mean is that tasks created for the inference process of a node, applied to a message, are added to the task queue.

The steps illustrated in Fig. 6 consist mostly of backward inference. The backward inference begins at the query, `wft1`, and continues until some channel is opened which contains an `i-infer` or `u-infer` message. In this example, this happens first at `wft10`, in step 5b of the figure. Listed below are the details of the processing which

---

[7] A KR formalism can be seen simultaneously as graph-, frame-, and logic-based. In our system, we define caseframes which have a number of slots, corresponding to argument positions in the logical view. The names of those slots are what decide the arc labels. See [15] for more details.

[8] The relations used here are based upon entries in the Unified Verb Index [22].

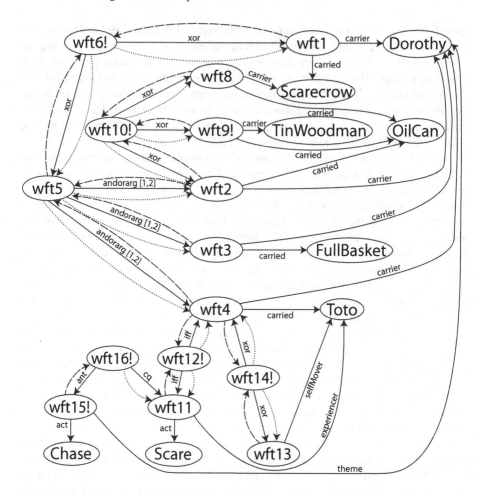

Fig. 5. The inference graph intended to mean that either Dorothy can carry the Scarecrow, or she can carry one or two of the following: Toto, her full basket, and the Tin Woodman's oil can; only one of the Scarecrow, the Tin Woodman, or Dorothy can carry the oil can; Dorothy carries Toto when, and only when, Toto is scared; Toto is either carried by Dorothy, or walks; if Dorothy is being chased, then Toto is scared; Dorothy is being chased; and the Tin Woodman is carrying his oil can

occurs during each step shown in the figure, along with the contents of the task queue. Tasks in the task queue are displayed in the following format:

$$\texttt{<wftSrc} -X \rightarrow \texttt{wftDest>}$$

where `wftSrc` is the source of the message which caused the creation of the task, `wftDest` is the node the task is operating within, and X is one of i,u,b, or c standing for the type of message the task processes, `i-infer`, `u-infer`, `backward-infer`, or `cancel-infer`. The `unassert` message type is not used in this example, and therefore was not listed.

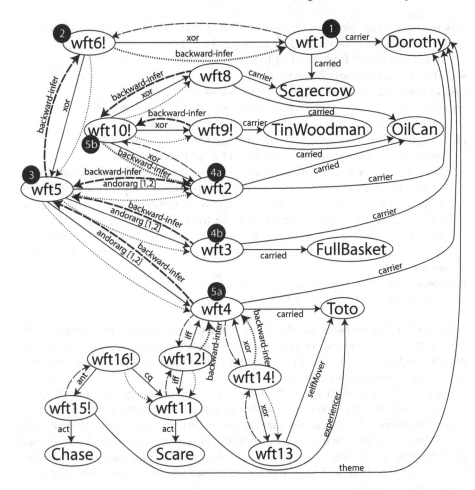

Fig. 6. The first five steps of inference when attempting to derive whether Dorothy is carrying the Scarecrow. Channels with a heavier weight and slightly larger arrows have had their channels opened through backward inference. Two processors are assumed to be used – a and b – and for this reason some steps in the graph have "a" or "b" appended to them. In these five steps, `backward-infer` messages flow backward through the graph until the first channel is reached with messages which will flow forward: the fact that `wft9` is asserted will flow to `wft10` since the i-channel connecting them has just been opened through backward inference.

1 `wft1` sends `backward-infer` message to `wft6!`;
    opens the channel from `wft6!` to `wft1`.
task queue $\langle$`wft1` $-b \rightarrow$ `wft6!`$\rangle$
2 `wft6!` sends `backward-infer` message to `wft5`;
    opens the channel from `wft5` to `wft6!`.
task queue $\langle$`wft6!` $-b \rightarrow$ `wft5`$\rangle$
3 `wft5` sends `backward-infer` messages to `wft2`, `wft3`, and `wft4`;
    opens the channels from `wft2`, `wft3`, and `wft4` to `wft5`.

task queue <wft5 $-b \to$ wft2>, <wft5 $-b \to$ wft3>, <wft5 $-b \to$ wft4>

4a wft2 sends backward-infer message to wft10!;
opens the channel from wft10! to wft4.

4b wft3 has no channels to open.

task queue <wft5 $-b \to$ wft4>, <wft2 $-b \to$ wft10!>

5a wft4 sends backward-infer messages to wft12! and wft14!;
opens the channels from wft12! and wft14! to wft4.

5b wft10! sends backward-infer messages to wft8 and wft9!;
opens the channels from wft8 and wft9! to wft10!.

Since wft9! is asserted, there is an i-infer message already waiting in the channel from wft9! to wft10! with higher priority than any backward inference tasks. That i-infer message is moved across the valve, and a new task is created for it – causing wft10! to be added to the front of the queue again. Since there was an i-infer message waiting at the opened valve from wft9! to wft10!, the backward-infer task just queued to occur in wft9! is canceled, as it is unnecessary.

task queue <wft9! $-i \to$ wft10!>, <wft4 $-b \to$ wft12!>,
<wft4 $-b \to$ wft14!>, <wft10! $-b \to$ wft8>

Remember that no backward-infer messages are sent to nodes which are already part of the derivation. For example, wft10 does not send a backward-infer message back to wft2 since wft2 is already part of the current derivation. This prevents eventual unnecessary derivations.

Figure 7 shows the next series of inference steps. In these steps the truth of wft9 is used to infer the negation of wft2, that Dorothy is not carrying the oil can (corresponding to step 3 in the earlier manual derivation), and relays this information to wft5. Backward inference continues from wft14 and wft12 back to wft13 and wft15, respectively. This is, again, the point where an i-infer message is ready to flow forward, this time from wft15 to wft16. Below we have once again described these processing steps in detail.

6a wft10! receives i-infer message from wft9!;
derives that both wft2 and wft8 are negated, by the rules of xor;
sends u-infer messages to both wft2 and wft8 telling them they are negated (of which, only the message to wft2 will pass through an open channel);
cancels any inference in progress or queued to derive wft8, since it is the only antecedent still attempting to satisfy wft10!.

6b wft12! sends backward-infer message to wft11;
opens the channel from wft11 to wft12!.

task queue <wft10! $-c \to$ wft8>, <wft10! $-u \to$ wft2>,
<wft4 $-b \to$ wft14!>, <wft12! $-b \to$ wft11>

7a wft8 has no channels to close.

7b wft2 receives u-infer message from wft10!;
asserts that it itself is negated (wft17!);
sends an i-infer message along the channel to wft5, telling wft5 that wft2 has been derived to be false.

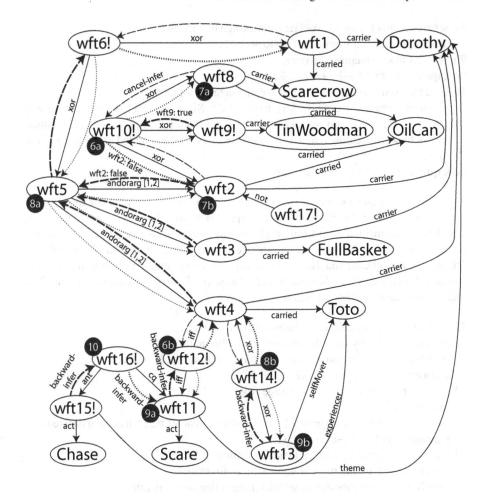

Fig. 7. Steps six through ten of the attempted derivation of whether Dorothy is carrying the Scarecrow. In steps 6b, 8b, 9a, 9b, and 10 backward inference is performed until the i-infer message indicating wft15 is true might flow forward across it's i-channel to wft16. Additionally, wft10 receives an i-infer message about the truth of wft9 (step 6a), which derives that wft2 is false through xor-elimination. wft2 then reports this (step 7b) to wft5, which records this information (step 8a), but cannot yet do anything else.

task queue <wft2 −*i* → wft5>, <wft4 −*b* → wft14!>,
    <wft12! −*b* → wft11>

8a wft5 receives i-infer message from (the negated) wft2. Since wft5 requires more information to determine if between 1 and 2 of its arguments are true, no more can be done.

8b wft14! sends backward-infer message to wft13;
    opens the channel from wft13 to wft14!.

task queue <wft12! −*b* → wft11>, <wft14! −*b* → wft13>

9a wft11 sends backward-infer message to wft16!;
  opens the channel from wft16! to wft11.
9b wft13 has no channels to open.
task queue <wft11 $-b \rightarrow$ wft16!>
10 wft16! sends backward-infer messages to wft15!;
  opens the channel from wft15! to wft16!. Since wft15! is asserted, there is
  an i-infer message waiting in the channel from wft15! to wft16!, which
  flows forward creating a new task to operate on wft16!, added at the front of the
  queue.
task queue <wft15! $-i \rightarrow$ wft16!>

Figure 8 illustrates the conclusion of the derivation that Dorothy is not carrying the
Scarecrow. In this final series of inference steps, it is derived that Toto is scared (wft11,
step 1 from the manual derivation), that Dorothy carries Toto (wft4, step 2 from the
manual derivation), that Dorothy carries one or two of the items from the list of: her
full basket, the oil can, and Toto (wft5, step 4 from the manual derivation), and that
Dorothy does not carry the Scarecrow (wft18, step 5 from the manual derivation). The
final set of inference steps are below.

11 wft16! receives i-infer message from wft15!;
  derives wft11 by the rules of entailment elimination;
  sends u-infer message to wft11 telling it that wft11 is true.
task queue <wft16! $-u \rightarrow$ wft11>
12 wft11 receives u-infer message from wft16!;
  asserts itself;
  sends i-infer message to wft12! telling it that wft11! has been derived.
task queue <wft11! $-i \rightarrow$ wft12!>
13 wft12! receives i-infer message from wft11!;
  derives wft4 by the rules of equivalence elimination;
  sends u-infer message to wft4 telling it that wft4 is true.
task queue <wft12! $-u \rightarrow$ wft4>
14 wft4 receives the u-infer message from wft12!;
  asserts itself;
  sends i-infer message to wft5 telling it that wft4! has been derived;
  sends a cancel-infer message to wft14!.
task queue <wft4! $-c \rightarrow$ wft14!>, <wft4! $-i \rightarrow$ wft5>
15a wft14! receives cancel-infer message from wft4!;
  sends a cancel-infer message to wft13.
15b wft5 receives the i-infer message from wft4!;
  derives itself through andor introduction, since between 1 and 2 of its antecedents
  must be true;
  sends an i-infer message to wft6!. sends a cancel-infer message to
  wft3.
task queue <wft5! $-c \rightarrow$ wft3>, <wft14! $-c \rightarrow$ wft13>,
  <wft5! $-i \rightarrow$ wft6!>
16a wft3 has no channels to close.

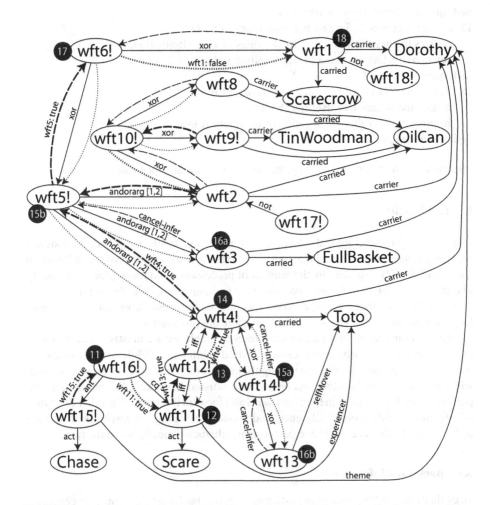

Fig. 8. The conclusion of the derivation that Dorothy is not, in fact, carrying the Scarecrow. Since wft15 is true (the fact that Dorothy is being chased), it is reported to wft16 (step 11), which fires and sends u-infer messages to its consequent – wft11. wft11, in step 12, reports its new truth value to wft12 which fires (step 13), since it is an if-and-only-if, and sends a u-infer message to wft4. Since wft4 is now true, it cancels other inference attempting to derive it and reports its truth to wft5 (step 14). Simultaneously, wft14 continues canceling inference (step 15a), and wft5 is found to be true, since it required between 1 and 2 of its antecedents to hold, and that is now the case (step 15b). Unnecessary inference attempting to derive wft5 and wft14 are canceled in steps 16a and b. Now, in step 17, wft6 receives the message that wft5 is true (that is, Dorothy is carrying 1 or 2 items), and because it is exclusive or, it determines that wft1 is false - Dorothy is not carrying the Scarecrow. Step 18 asserts this fact, and reports it to the user.

16b `wft13` has no channels to close.

task queue `<wft5! −i → wft6!>`

17 `wft6!` receives `i-infer` message from `wft5!`;

derives the negation of `wft1` by the rules of `xor` elimination;

sends `u-infer` message to `wft1` telling it that `wft1` is false.

task queue `<wft6! −u → wft1>`

18 `wft1` receives `u-infer` message from `wft6!`;

asserts the negation of itself (`wft18!`);

informs the user that the negation of the query is true.

task queue *empty*

Inference is now complete, and any channels which remain open can be closed.

6   Evaluation

The massively parallel logical inference systems of the 1980s and 90s often assigned each processor a single formula or rule to be concerned with. This resulted in limits on the size of the KB (bounded by the number of processors), and many processors sitting idle during any given inference process. Our technique dynamically assigns tasks to threads only when they have work to do, meaning that processors are not sitting idle so long as there are as many tasks available as there are processors.

In evaluating the performance of the inference graph, we are mostly concerned with the speedup achieved as more processors are used in inference. While overall processing time is also important, if speedup is roughly linear with the number of processors, that will show that the architecture and heuristics discussed scale well. We will look at the performance of our system in both backward and forward inference. The other two types of inference – bi-directional inference, and focused reasoning – are hybrids of forward and backward inference, and have performance characteristics between the two.

6.1   Backward Inference

To evaluate the performance of the inference graph in backward inference, we generated graphs of chaining entailments. Each entailment had bf antecedents, where bf is the branching factor, and a single consequent. Each consequent was the consequent of exactly one rule, and each antecedent was the consequent of another rule, up to a depth of d entailment rules. Exactly one consequent, cq, was not the antecedent of another rule. Therefore there were bf^d entailment rules, and $2*bf^d-1$ antecedents/consequents. Each of the bf^d leaf nodes were asserted. We tested the ability of the system to backchain on and derive cq when the entailments used were both and-entailment and or-entailment. Backward inference is the most resource intensive type of inference the inference graphs can perform, and most fully utilizes the scheduling heuristics developed in this paper.

In the first test we used and-entailment, meaning for each implication to derive its consequent both its antecedents had to be true. Since we backchained on cq, this meant

Table 1. Inference times using 1, 2, 4, and 8 CPUs for 100 iterations of and-entailment in an inference graph with $d = 10$ and $bf = 2$ in which all 1023 rule nodes must be used to infer the result

CPUs	Inference Time (ms)	Speedup
1	37822	1.00
2	23101	1.64
4	15871	2.39
8	11818	3.20

every node in the graph would have to become asserted. This is the worst case scenario for entailment. The timings we observed are presented in Table 1.[9]

In increasing the number of usable CPUs from 1 to 2, we achieve nearly double the performance. As we increase further, there is still a benefit, but the advantage begins to drop off. The primary reason for this is that 1023 formulas must be asserted to the KB, and this requires maintenance of shared state. Only one thread can modify the shared state at a time, and so we handle it synchronously (as explained in Sect. 4). We found 100 iterations of asserting 1023 formulas which had already been built, as in the case of this test, took approximately 7500ms, regardless of the number of CPUs used. Excluding this from each of the times in Table 1 reveals a close-to-halving trend every time the number of CPUs is doubled, as would be expected (See Table 2).

Table 2. The results from Table 1, excluding the time (7500ms) for assertions

CPUs	Inference−Assert Time (ms)	Speedup
1	30322	1.00
2	15601	1.94
4	8371	3.62
8	4318	7.02

We then tried to determine whether the depth or branching factor of the graph has any effect on speedup as the number of processors increases. We first ran an experiment to judge the impact of graph depth. We ran the experiment on graphs of five different depths, ranging from 5 to 15 (32 leaves, to 32,768 leaves), while maintaining $bf = 2$ (see Fig. 9), and found that as graph depth is increased, speedup increases very slowly. Since this increase must be bounded (we have no reason to believe a more-than-doubling speedup is possible), we have fit logarithmic trendlines to the 2, 4, and 8 CPU data, and found R^2 values to suggest a strong fit.

To find out if branching factor affects speedup, we chose $d = 7$ (128 leaves, 127 rule nodes), and varied the branching factor from 1 to 4. When $bf = 1$, the graph is simply

---

[9] All tests were performed on a Dell Poweredge 1950 server with dual quad-core Intel Xeon X5365 processors (no Hyper-Threading) and 32GB RAM. Each test was performed twice, with the second result being the one used here. The first run was only to allow the JVM to "warm up."

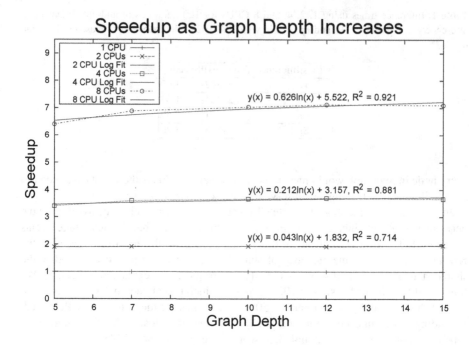

Fig. 9. Speedup of the and-entailment test, shown in relation to the depth of the inference graph. As the depth of the graph increases, speedup increases slowly but logarithmically with each number of CPUs tested. A branching factor of 2 was used in all tests.

a chain of nodes, and the use of more processors can provide no possible improvement in computation time. In fact, as shown in Fig. 10, throwing more processors at the problem makes things worse. Fortunately, this is a rather contrived use of the inference graph. At branching factors 2-4, the graph performs as expected, with the branching factor increase having little effect on performance.[10] There may be a slight performance impact as the branching factor increases, because the RUI computations happening in the rule nodes rely on shared state, but that performance hit only occurs when two tasks attempt to modify the RUI set of a single node simultaneously – an increasingly unlikely event as we consider larger graphs.

In our second test we used the same KB from the first ($d = 10$, $bf = 2$), except each and-entailment rule was swapped for or-entailment. Whereas the earlier test required every consequent in the KB to be derived, this test shows the best case of entailment - only a single leaf must be found to be asserted to allow the chaining causing cq to become true.

The improvement in or-entailment processing times as we add more CPUs (see Table 3) is not as dramatic since the inference operation performed once a chain of valves from an asserted formula to cq are open cannot be accelerated by adding more processing

---

[10] Because of the graph size explosion as branching factor increases, it was not possible to collect enough data to perform a quantitative analysis, only a qualitative one.

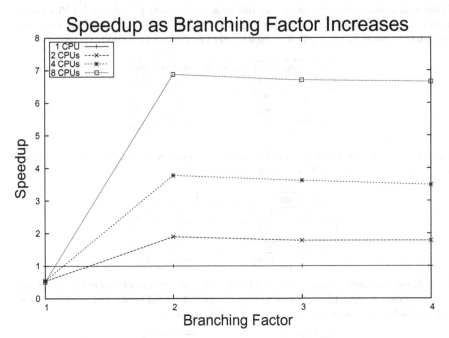

Fig. 10. Speedup of the and-entailment test, shown in relation to the branching factor of the inference graph. A depth of 7 was used in all tests, to keep the number of nodes reasonable.

cores – that process is inherently sequential. By timing forward inference through the graph, we found that approximately 340ms in each of the times in Table 3 is attributed to the relevant inference task. The improvement we see increasing from 1 to 2, and 2 to 4 CPUs is because the `backward-infer` and `cancel-infer` messages spread throughout the network and can be sped up through concurrency. During the periods where backward inference has already begun, and `cancel-infer` messages are not being sent, the extra CPUs were working on deriving formulas relevant to the current query, but in the end unnecessary – as seen in the number of rule nodes fired in Table 3. The time required to assert these extra derivations begins to outpace the improvement gained through concurrency when we reach 8 CPUs in this task. These extra derivations may be used in future inference tasks, though, without re-derivation, so the resources are not wasted. As only a chain of rules are required, altering branching factor or depth has no effect on speedup.

The difference in computation times between the or-entailment and and-entailment experiments are largely due to the scheduling heuristics described in Sect. 4.1. Without the scheduling heuristics, backward inference tasks continue to get executed even once messages start flowing forward from the leaves. Additionally, more rule nodes fire than is necessary, even in the single processor case. We ran the or-entailment test again without these heuristics using a FIFO queue, and found the inference took just as long as in Table 1. We then tested a LIFO queue since it has some of the characteristics of our prioritization scheme (see Table 4), and found our prioritization scheme to be nearly 10x faster.

Table 3. Inference times, and number of rule nodes used, using 1, 2, 4, and 8 CPUs for 100 iterations of or-entailment in an inference graph ($d = 10$, $bf = 2$) in which there are many paths through the network (all of length 10) which could be used to infer the result

CPUs	Time (ms)	Avg. Rules Fired	Speedup
1	1021	10	1.00
2	657	19	1.55
4	498	38	2.05
8	525	67	1.94

Table 4. The same experiment as Table 3 replacing the improvements discussed in Sect. 4.1, with a LIFO queue. Our results in Table 3 are nearly 10x faster.

CPUs	Time (ms)	Avg. Rules Fired	Speedup
1	12722	14	1.00
2	7327	35	1.77
4	4965	54	2.56
8	3628	103	3.51

Table 5. Inference times using 1, 2, 4, and 8 CPUs for 100 iterations of forward inference in an inference graph of depth 10 and branching factor 2 in which all 1024 leaf nodes are derived. Each result excludes 7500ms for assertions, as discussed in Sect. 6.1.

CPUs	Inference−Assert Time (ms)	Speedup
1	30548	1.00
2	15821	1.93
4	8234	3.71
8	4123	7.41

6.2   Forward Inference

To evaluate the performance of the inference graph in forward inference, we again generated graphs of chaining entailments. Each entailment had a single antecedent, and 2 consequents. Each consequent was the consequent of exactly one rule, and each antecedent was the consequent of another rule, up to a depth of 10 entailment rules. Exactly one antecedent, *ant*, the "root", was not the consequent of another rule. There were 1024 consequents which were not antecedents of other rules, the leaves. We tested the ability of the system to derive the leaves when *ant* was asserted with forward inference.

Since all inference in our graphs is essentially forward inference (modulo additional message passing to manage the valves), and we're deriving every leaf node, we expect to see similar results to the and-entailment case of backward inference, and we do, as shown in Table 5. The similarity between these results and Table 2 shows the relatively small impact of sending `backward-infer` messages, and dealing with the shared state in the RUIs. Excluding the assertion time as discussed earlier, we again show a near doubling of speedup as more processors are added to the system. In fact, altering the branching factor and depth also result in speedups very similar to those from Figs. 9 and 10.

7   Conclusions

Inference graphs are an extension of propositional graphs capable of performing natural deduction using forward, backward, bi-directional, and focused reasoning within a concurrent processing system. Inference graphs add channels to propositional graphs, built along the already existing edges. Channels carry prioritized messages through the graph for performing and controlling inference. The priorities of messages influence the order in which tasks are executed – ensuring that the inference tasks which can derive the user's query most quickly are executed first, and irrelevant inference tasks are canceled. The heuristics developed in this paper for prioritizing and scheduling the execution of inference tasks improve performance in backward inference with or-entailment nearly 10x over just using LIFO queues, and 20-40x over FIFO queues. In and-entailment using backward inference, and in forward inference, our system shows a near linear performance improvement with the number of processors (ignoring the intrinsically sequential portions of inference), regardless of the depth or branching factor of the graph.

Acknowledgments. This work has been supported by a Multidisciplinary University Research Initiative (MURI) grant (Number W911NF-09-1-0392) for Unified Research on Network-based Hard/Soft Information Fusion, issued by the US Army Research Office (ARO) under the program management of Dr. John Lavery. We gratefully appreciate this support.

References

1. Baum, L.F.: The Wonderful Wizard of Oz. G. M. Hill (1900)
2. Choi, J., Shapiro, S.C.: Efficient implementation of non-standard connectives and quantifiers in deductive reasoning systems. In: Proceedings of the Twenty-Fifth Hawaii International Conference on System Sciences, pp. 381–390. IEEE Computer Society Press, Los Alamitos (1992)
3. Dixon, M., de Kleer, J.: Massively parallel assumption-based truth maintenance. In: Reinfrank, M., Ginsberg, M.L., de Kleer, J., Sandewall, E. (eds.) Non-Monotonic Reasoning 1988. LNCS, vol. 346, pp. 131–142. Springer, Heidelberg (1988)
4. Doyle, J.: A truth maintenance system. Artificial Intelligence 19, 231–272 (1979)
5. Forgy, C.: Rete: A fast algorithm for the many pattern/many object pattern match problem. Artificial Intelligence 19, 17–37 (1982)
6. Hickey, R.: The Clojure programming language. In: Proceedings of the 2008 Symposium on Dynamic Languages. ACM, New York (2008)
7. Huang, S.S., Green, T.J., Loo, B.T.: Datalog and emerging applications: an interactive tutorial. In: Proceedings of the 2011 ACM SIGMOD International Conference on Management of Data, SIGMOD 2011, pp. 1213–1216. ACM, New York (2011)
8. de Kleer, J.: Problem solving with the ATMS. Artificial Intelligence 28(2), 197–224 (1986)
9. Lehmann, F. (ed.): Semantic Networks in Artificial Intelligence. Pergamon Press, Oxford (1992)
10. Lendaris, G.G.: Representing conceptual graphs for parallel processing. In: Conceptual Graphs Workshop (1988)
11. Martins, J.P., Shapiro, S.C.: A model for belief revision. Artificial Intelligence 35, 25–79 (1988)

12. McAllester, D.: Truth maintenance. In: Proceedings of the Eighth National Conference on Artificial Intelligence (AAAI 1990), Boston, MA, pp. 1109–1116 (1990)
13. McKay, D.P., Shapiro, S.C.: Using active connection graphs for reasoning with recursive rules. In: Proceedings of the Seventh International Joint Conference on Artificial Intelligence, pp. 368–374. Morgan Kaufmann, Los Altos (1981)
14. Schlegel, D.R.: Concurrent inference graphs (doctoral consortium abstract). In: Proceedings of the Twenty-Seventh AAAI Conference (AAAI 2013), pp. 1680–1681 (2013)
15. Schlegel, D.R., Shapiro, S.C.: Visually interacting with a knowledge base using frames, logic, and propositional graphs. In: Croitoru, M., Rudolph, S., Wilson, N., Howse, J., Corby, O. (eds.) GKR 2011. LNCS, vol. 7205, pp. 188–207. Springer, Heidelberg (2012)
16. Schlegel, D.R., Shapiro, S.C.: Concurrent reasoning with inference graphs (student abstract). In: Proceedings of the Twenty-Seventh AAAI Conference (AAAI 2013), pp. 1637–1638 (2013)
17. Shapiro, E.: The family of concurrent logic programming languages. ACM Comput. Surv. 21(3), 413–510 (1989)
18. Shapiro, S.C.: Set-oriented logical connectives: Syntax and semantics. In: Lin, F., Sattler, U., Truszczynski, M. (eds.) Proceedings of the Twelfth International Conference on the Principles of Knowledge Representation and Reasoning (KR 2010), pp. 593–595. AAAI Press, Menlo Park (2010)
19. Shapiro, S.C., Martins, J.P., McKay, D.P.: Bi-directional inference. In: Proceedings of the Fourth Annual Conference of the Cognitive Science Society, pp. 90–93. The Program in Cognitive Science of The University of Chicago and The University of Michigan, Ann Arbor, MI (1982)
20. Shapiro, S.C., Rapaport, W.J.: The SNePS family. Computers & Mathematics with Applications 23(2-5), 243–275 (1992), reprinted in [9, pp. 243–275]
21. The Joint Task Force on Computing Curricula, Association for Computing Machinery, IEEE-Computer Society: Computer Science Curricula 2013 (2013)
22. University of Colorodo: Unified verb index (2012), http://verbs.colorado.edu/verb-index/index.php
23. Wachter, M., Haenni, R.: Propositional DAGs: a new graph-based language for representing boolean functions. In: KR 2006, 10th International Conference on Principles of Knowledge Representation and Reasoning, pp. 277–285. AAAI Press, U.K. (2006)
24. Yan, F., Xu, N., Qi, Y.: Parallel inference for latent dirichlet allocation on graphics processing units. In: Proceedings of NIPS, pp. 2134–2142 (2009)

Formal Concept Analysis
over Graphs and Hypergraphs

John G. Stell

School of Computing, University of Leeds, U.K.

Abstract. Formal Concept Analysis (FCA) provides an account of classification based on a binary relation between two sets. These two sets contain the objects and attributes (or properties) under consideration. In this paper I propose a generalization of formal concept analysis based on binary relations between hypergraphs, and more generally between pre-orders. A binary relation between any two sets already provides a bipartite graph, and this is a well-known perspective in FCA. However the use of graphs here is quite different as it corresponds to imposing extra structure on the sets of objects and of attributes. In the case of objects the resulting theory should provide a knowledge representation technique for structured collections of objects. The generalization is achieved by an application of work on mathematical morphology for hypergraphs.

1 General Introduction

1.1 Formal Concept Analysis

We recall the basic notions of Formal Concept Analysis from [GW99] with some notation taken from [DP11].

Definition 1. *A **Formal Context** $\mathbb{K}$ consists of two sets U, V and a binary relation $R \subseteq U \times V$. The elements of U are the **objects** of $\mathbb{K}$ and the elements of V are the **properties**.*

The relation can be written in the infix fashion $a \, R \, b$ or alternatively as $(u, v) \in R$. The converse of R is denoted $\check{R}$.

Definition 2. *Given a formal context (U, R, V) then $R^{\vartriangle} : \mathscr{P}U \to \mathscr{P}V$ is defined for any $X \subseteq U$ by*

$$R^{\vartriangle}(X) = \{v \in V : \forall x \in X \, (x \, R \, v)\}.$$

This operator provides the set of properties which every one of the objects in X possesses. Using the converse of R, the operator $\check{R}^{\vartriangle} : \mathscr{P}V \to \mathscr{P}U$ can be described explicitly by

$$\check{R}^{\vartriangle}(Y) = \{u \in U : \forall y \in Y \, (u \, R \, y)\}$$

for any $Y \subseteq V$.

Definition 3. *A **formal concept** belonging to the context (U, R, V) is a pair (X, Y) where $X \subseteq U$, $Y \subseteq V$ and such that $R^{\vartriangle}(X) = Y$ and $\check{R}^{\vartriangle}(Y) = X$. The sets X and Y are respectively the **extent** and the **intent** of the concept.*

M. Croitoru et al. (Eds.): GKR 2013, LNAI 8323, pp. 165–179, 2014.
© Springer International Publishing Switzerland 2014

1.2   Importance of Graphs and Hypergraphs

The question posed in this paper is what happens when we replace the sets U and V in a formal context by graphs, or by hypergraphs or more generally still by pre-orders? The question is not straightforward, partly because there is more than one possible meaning for a relation between graphs. First, however, we need to consider why the question might be worth answering.

The idea that properties can have structure is well established within FCA, see for example the treatment of scales and many-valued contexts in [GW99, p36]. Suppose however our collection of objects also has structure (rather than each individual object being structured).

We now consider some examples.

Assemblages. An assemblage of individuals where the individuals have independent existence. For example, a marriage between two people or a group of four friends who take a holiday together. Further examples would be alliances of nations, twinnings of towns. In all these cases we can consider attributes of the individuals and also attributes of the groupings. A particular grouping can have attributes which may not depend solely on the individuals, and the same set of individuals may participate in a number of different groupings.

Granules. Such groupings could also appear in granulation, equivalence classes are a simple case but other kinds of granules are also well known. A granule can have attributes (say some viewpoint that engenders the clustering) and again the same individuals may appear in different granules for different viewpoints. This indicates the need to attach attributes to the granules themselves as well as to the individuals.

Links. In a different kind of example the collection of objects could make up both the stations and lengths of track joining stations to be found in a railway system. Similarly there might be cities linked by roads or nodes and links in a social network. In the case of a railway network it is clear that the railway line between two stations can require different attributes from the stations themselves.

All of these examples can be modelled as graphs or hypergraphs, and we shall restrict to undirected graphs in the present paper.

2   Theoretical Background

2.1   Graphs and Hypergraphs

A hypergraph can be defined as consisting of a set N of nodes and a set E of edges (or hyperedges) together with an incidence function $i : E \to \mathscr{P}N$ from E to the powerset of N. This approach allows several edges to be incident with the same set of nodes, and also allows edges incident with the empty set of nodes. An example is shown in Figure 1 where the edges are drawn as curves enclosing the nodes with which they are incident.

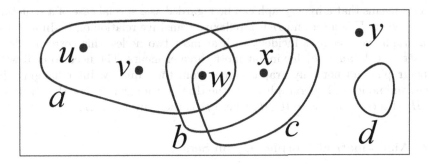

Fig. 1. Hypergraph with edges $E = \{a, b, c, d\}$, and nodes $N = \{u, v, w, x, y\}$

When studying relations on these structures it is more convenient to use an equivalent definition, in which there is a single set comprising both the edges and nodes together. This has been used in [Ste07, Ste10] and is based on using a similar approach to graphs in [BMSW08]. This is illustrated in Figure 2 and set out formally in Definition 4.

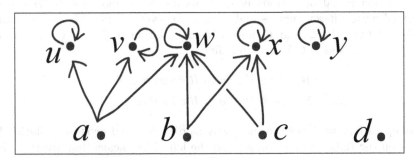

Fig. 2. Hypergraph from Figure 1 represented as a relation φ on the set U of all edges and nodes

Definition 4. *A **hypergraph** consists of a set U and a relation $\varphi \subseteq U \times U$ such that for all $x, y, z \in U$,*

1. *if $(x, y) \in \varphi$ then $(y, y) \in \varphi$, and*
2. *if $(x, y) \in \varphi$ and $(y, z) \in \varphi$ then $y = z$.*

From a hypergraph described in this way we can obtain the edges as those $u \in U$ for which $(u, u) \notin \varphi$, whereas the nodes satisfy $(u, u) \in \varphi$. A **sub-hypergraph** of (U, φ) is defined as a subset $K \subseteq U$ for which $k \in K$ and $(k, u) \in \varphi$ imply $u \in K$. We will usually say sub-graph rather than sub-hypergraph for brevity.

It is technically convenient to replace the relation φ by its reflexive closure $\varphi^\circlearrowright$. This loses some information, as we cannot distinguish between nodes incident with no edges and edges incident with no nodes. However, this loss is not significant as Stell [Ste12] shows that the relations we need to define on hypergraphs are the same in each case.

This means that a hypergraph can be regarded as a special case of a structure (U, H) where U is a set and H is a reflexive transitive relation on U. In the case of a graph every edge is incident with at most two nodes, this corresponds to the set $\{v \in U : u \, H \, v\}$ having at most three elements. The need to deal with hypergraphs and not only graphs comes from their duality. Interchanging the edges and nodes in hypergraph yields the **dual hypergraph**. For the structure (U, H) this corresponds to $(U, \breve{H})$ where $\breve{H}$ is the converse of H.

2.2 Mathematical Morphology on Sets

Mathematical morphology [Ser82, BHR07] has been used widely in image processing and has an algebraic basis in lattice theory. It can be seen as a way of approximating images so as to emphasise essential features. At the most abstract level it depends on the action of binary relations on subsets. Although various authors [CL05, ZS06, DP11] have analysed several connections between different approaches to granulation and approximation, the exploration of the links between mathematical morphology and formal concept analysis is only beginning to be developed [Blo11].

The fundamental operations used in mathematical morphology have been described with different names and notions in several fields. The basic setting consists of two sets, U and V, and a binary relation $R \subseteq U \times V$. Given any subsets, $X \subseteq U$ and $Y \subseteq V$ we can define

$$X \oplus R = \{v \in V : \exists u \, (u \, R \, v \text{ and } u \in X)\}$$
$$R \ominus Y = \{u \in U : \forall v \, (u \, R \, v \text{ implies } v \in Y)\}$$

These operations are known respectively as **dilation** and **erosion**. Dilation by R acts on the right and erosion acts on the left. This means that given further relations composable with R, say $S \subseteq U' \times U$ and $T \subseteq V \times V'$ then $X \oplus (R \,;\, T) = (X \oplus R) \oplus T$ and $(S \,;\, R) \ominus Y = S \ominus (R \ominus Y)$.

The basic image processing context of these constructions is usually that the sets U and V are both $\mathbb{Z}^2$ thought of as a grid of pixels and the relation R is generated by a particular pattern of pixels. Extensions of the operations to graphs have been discussed from various viewpoints [HV93, CNS09], and approaches to morphology on hypergraphs include [Ste10, BB11].

The operation $R^\triangle$, introduced in Definition 2, is related to dilation and erosion as follows. In these constructions $\hat{R}$ is the converse complement of R, that is $\hat{R} = \breve{\overline{R}} = \overline{\breve{R}}$, where $\overline{R}$ denotes the complement of the relation R. The complement of a set X is denoted $-X$. The proof is a straightforward calculation from the definitions.

Theorem 1. *Let $R \subseteq U \times V$ be a relation and $X \subseteq U$ and $Y \subseteq V$. Then*

$$R^\triangle(X) = \hat{R} \ominus (-X),$$
$$\breve{R}^\triangle(Y) = -(Y \oplus \hat{R}).$$

The reason for establishing this result is that it provides a guide for generalizing formal concept analysis from sets to graphs including the potential of extending the treatment of decomposition of contexts in [DP11]. It also provides a foundation for developing correspondences between formal concept analysis and mathematical morphology. This is clear once the operations of (morphological) opening and closing, which are derived from combinations of dilation and erosion are introduced.

Corollary 2. (X, Y) *is a formal concept iff*

$$-X = (\hat{R} \ominus -X) \oplus \hat{R}, \text{ and}$$
$$Y = \hat{R} \ominus (Y \oplus \hat{R}).$$

In other words, $-X$ is open (equal to its own morphological opening) and Y is closed (equal to its own morphological closing) with respect to the relation $\hat{R}$.

The calculus of erosion and dilation makes particularly clear the well-known fact that the intent of a concept determines the extent and vice versa. Given Y where $Y = \hat{R} \ominus (Y \oplus \hat{R})$ we can define $X = -(Y \oplus \hat{R})$ and immediately obtain $Y = \hat{R} \ominus (-X)$. Similarly if we are just given the extent X We shall see in Section 3 how the theory of mathematical morphology on hypergraphs allows us to formulate a generalization of formal concept analysis.

2.3   The Algebra of Graphs and Hypergraphs

Given a graph, by which we mean an undirected multigraph with possible loops, the set of all subgraphs forms a bi-Heyting algebra, generalizing the Boolean algebra of subsets of a set. This bi-Heyting structure has been described in [Law86, RZ96], but we review it here, in the more general setting of hypergraphs, as it provides motivation for later sections.

Given a hypergraph (U, H), a **sub-graph** is any $X \subseteq U$ for which $X \oplus H \subseteq X$. The sub-graphs form a complete sub-lattice $\mathscr{L}$ of the lattice of all subsets of U. The meet and join operations in $\mathscr{L}$ are the usual intersections and meets of subsets. If X is a sub-hypergraph, then its complement $U - X$ is not necessarily a sub-hypergraph. There are, however, two weaker operations on sub-hypergraphs: the pseudocomplement, $\neg X$, and the dual pseudocomplement, $\dashv X$. These satisfy both $X \cup \neg X = U$ and $X \cap \neg X = \varnothing$ but not necessarily $X \cup \neg X = U$ or $X \cap \dashv X = \varnothing$.

These can be expressed using dilation and erosion with respect to H:

$$\neg X = H \ominus (-X),$$
$$\dashv X = (-X) \oplus H.$$

These two weaker forms of the usual Boolean complement provide an expressive language allowing us to define notions such as boundary, expansion and contraction for subgraphs. These are illustrated in the following diagrams where we have taken a subgraph X of a graph for simplicity, but the constructions have similar interpretations for subgraphs of a hypergraph.

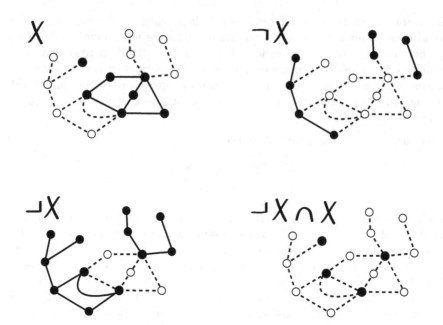

The **boundary** of X consists of those nodes incident with some edge which is not in X. The boundary is $\neg X \cap X$.

The **expansion** of X contains all of X together with any edges incident with a node in X and any nodes incident with these edges. Intuitively, this is X expanded by anywhere accessible along an edge from X. The expansion is $\neg\neg X$. In the other direction the **contraction** contains only the nodes which are not incident with anything in the boundary, and it also contains any edges in X which are exclusively incident with these nodes.

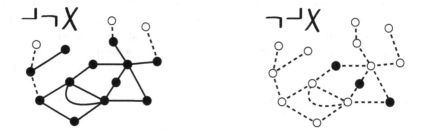

The other two combinations of the operations, in which one of the weaker converses is repeated twice, provide the **node opening**, $\neg\neg X$, and the **edge closing**, $\neg\neg X$. The node opening removes from X any nodes which are incident only with edges which are not in X. The edge closing adds to X any edges all of whose incident nodes are in X.

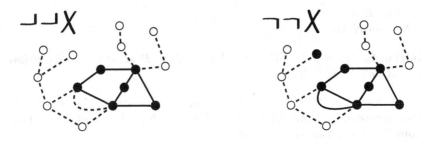

When a subgraph X forms the extent of a formal concept the availability of the four, potentially distinct (as in the above example), subgraphs: $\neg\dashv X$, $\neg\neg X$, $\dashv\neg X$, $\dashv\dashv X$ can be expected to produce a much richer theory than the classical when X is a set. In this classical case of course all four of these reduce to $--X$ which is just X itself.

3   Contexts and Concepts for Pre-orders

3.1   Relations between Preorders

A formal context is a relation between two sets. To generalize this to relations between graphs or hypergraphs we have to consider what should be meant by a relation between such structures. We find it is convenient to work in the more general setting of pre-orders, but it is especially useful to work with examples where the pre-order defines the incidence structure in a graph or a hypergraph.

There are several different possibilities for how to define a relation between graphs. As relations are subsets of cartesian products we might try defining a relation between graphs as a subgraph of a product of graphs, where 'product of graphs' would need to be understood in a suitable category of graphs. While it is straightforward to define products of directed graphs, the undirected case is more subtle [BL86, BMSW08] requiring some adjustment to the usual notions of undirected graph and of morphism between such structures. Instead of developing this approach, we use the notion studied in [Ste12] which generalizes the characterization of relations R between sets U and V with $R \subseteq U \times V$ as union-preserving functions from $\mathscr{P}U$ to $\mathscr{P}V$. This means we use a notion of relation between hypergraphs, (U, H) and (V, K), equivalent to union-preserving functions from the lattice of subgraphs of (U, H) to the lattice of subgraphs of (V, K).

We assume in this section that U and V are sets and that $H \subseteq U \times U$ and $K \subseteq V \times V$ are both reflexive and transitive relations (often called preorders). The appropriate generalization of subgraph to this situation is as follows.

Definition 5. *If $X \subseteq U$ is any subset, then X is a H-set if $X \oplus H \subseteq X$.*

The lattice of H-sets on (U, H) will be denoted $\mathscr{L}(U, H)$.

Proposition 3. *The following are equivalent*

1. X *is an* H*-set.*
2. $-X$ *is an* $\check{H}$*-set*

Definition 6. *Let* $R \subseteq U \times V$ *be a relation. Then* R *is an* (H, K)*-relation if* $H \,;\, R \subseteq R$ *and* $R \,;\, K \subseteq R$.

This extends the notion of a relation on a hypergraph studied in [Ste12] which is a special case of the situation when $(U, H) = (V, K)$. The significance of these relations lies in the following property.

Theorem 4. *There is a bijection between the set of* (H, K)*-relations and the set of join-preserving functions from* $\mathscr{L}(U, H)$ *to* $\mathscr{L}(V, K)$. $\qquad\square$

The result is proved by a straightforward generalization of the well-known case of H and V just being the identity relations on U and V respectively (that is when R is an ordinary relation between sets). The bijection preserves both the compositional structure and the way relations act on H-sets in U. The appropriate setting for the correspondence would be in terms of quantaloids [Ros96] but this is beyond the scope of this paper.

For the applications to formal concept analysis, we need to consider relations which are $(\check{H}, K)$-relations, while also considering H-sets rather than $\check{H}$-sets in U. It is straightforward to show that the converse of an $(\check{H}, K)$-relation is not necessarily a $(K, \check{H})$-relation, and the complement of an $(\check{H}, K)$-relation is not necessarily a $(\check{H}, K)$-relation. However, the operations of converse and complement do have the properties stated in Proposition 6. In order to establish these properties we need to recall the properties of the residuals for relations.

Definition 7. *Let* $Q \subseteq U \times V$ *and* $R \subseteq V \times W$ *be relations, and let* $S \subseteq U \times W$ *also be a relation.*
The **Left Residual**, $\backslash$, *is defined by* $Q \backslash S = \overline{\check{Q} \,;\, \overline{S}}$.
The **Right Residual**, $/$, *is defined by* $R \,/\, S = \overline{\overline{R} \,;\, \check{S}}$.

The following well-known property of these residuals is important.

$$Q \,;\, R \subseteq S \quad \text{iff} \quad R \subseteq Q \backslash S \quad \text{iff} \quad Q \subseteq R \,/\, S \qquad (1)$$

Lemma 5. *Let* $R \subseteq U \times V$ *and let* H *and* K *be pre-orders on* U *and* V *respectively. Then*

1. $H \,;\, \overline{R} \subseteq \overline{R}$ *iff* $\check{H} \,;\, R \subseteq R$,
2. $R \,;\, \check{K} \subseteq \overline{R}$ *iff* $R \,;\, K \subseteq R$. $\qquad\square$

Proof. We have $H \,;\, \overline{R} \subseteq \overline{R}$ iff $\overline{R} \subseteq H \backslash \overline{R}$ by (1). But we also have $\check{H} \,;\, R \subseteq R$ iff $\overline{R} \subseteq \check{H} \,;\, R = H \backslash \overline{R}$. This gives the first part, and the second is similar using the other residual.

Proposition 6. *The following are equivalent for $R \subseteq U \times V$.*

1. *R is an $(\breve{H}, K)$-relation.*

2. *$\overline{R}$ is an $(H, \breve{K})$-relation.*

3. *$\breve{R}$ is a $(\breve{K}, H)$-relation.*

4. *$\hat{R}$ is a $(K, \breve{H})$-relation.*

Proof. The first two parts are equivalent by Lemma 5. The first and third parts are equivalent by basic properties of the converse operation. Finally if R is an $(\breve{H}, K)$-relation then $\overline{R}$ is an $(H, \breve{K})$-relation, and $\overline{\breve{R}}$ is a $(\breve{K}, \breve{H})$-relation so the first part implies the fourth, and the fourth implies the first since the converse-complement is an involution.   □

Theorem 7. *If $X \subseteq U$ and $Y \subseteq V$ are any subsets and R is an (H, K)-relation then*

1. *$X \oplus R$ is a K-set, and*
2. *$R \ominus Y$ is an H-set.*

Proof. For the first part, we have the calculation

$$(X \oplus R) \oplus K = X \oplus (R \,;\, K) \subseteq X \oplus R.$$

For the second, suppose $x \in R \ominus Y$ and $x \, H \, x'$. If $x' \, R \, v$ where $v \in -Y$ then we have a contradiction as $x \, R \, v$.   □

Using a special case of this (when X is an H-set), it can be shown that dilation provides a union-preserving mapping from $\mathscr{L}(U, H)$ to $\mathscr{L}(V, K)$. This dilation then forms a Galois connection with the erosion of K-sets. This confirms that we have established an appropriate generalization of mathematical morphology on sets, and that we can re-capture the usual case by taking H and K to be the identity relations on U and V.

This generalization of mathematical morphology now allows us to use Theorem 1 and Corollary 2 as the basis of a generalization of formal concept analysis from sets of properties and objects to hypergraphs of properties and objects. More generally still the approach works for pre-orders with their associated H-sets as collections of objects and K-sets as collections of properties.

3.2   Defining Formal Concepts

Definition 8. *A **pre-order context** is a pair $((U, H), (V, K))$, where H and K are pre-orders on the sets U and V respectively, together with an $(\breve{H}, K)$ relation $R \subseteq U \times V$.*

It is sometimes convenient to denote a context as $R : (U, H) \to (V, K)$. Here the arrow is to be understood as indicating not a function between sets, but as a morphism in the category where the objects are pre-orders and the morphisms are relations with the appropriate properties.

Because the bi-Heyting algebra of H-sets is more complex than the Boolean algebra of all subsets of U, as noted at the end of Section 2.3 above, there is more than one possible definition of a formal concept in the generalized setting. These will need to be evaluated against potential applications, some of which are indicated in the following section.

The following is a corollary to Theorem 7 established using Proposition 6.

Corollary 8. *Suppose we have a formal context, with notation as in the above definition. If X and Y are respectively subsets of U and V then*

1. *$\hat{R} \ominus X$ is a K-set in (V, K),*
2. *$-(Y \oplus \hat{R})$ is an H-set in (U, H).*   □

This means that we can make the following definition for what should be meant by a formal concept when we have pre-orders and not merely sets.

Definition 9. *A **formal pre-order concept** in the formal context $R : (U, H) \to (V, K)$ is a pair (X, Y) where X is an H-set, Y is a K-set, and the following hold:*

$$X = -(Y \oplus \hat{R})$$
$$Y = \hat{R} \ominus (-X).$$

However, this is not the only possibility because both X and Y have some spatial structure. The condition on Y characterizes it as

'all those properties which only objects *outside* X do not have'.

In Definition 9 the *outside* of X is expressed by $-X$, but as Section 2.3 showed there are other notions of *outside*, namely $\neg X$ and $\dashv X$. The next result shows that the second of these does not lead to a different notion of intent.

Theorem 9. *If X is an H-set then*

$$\hat{R} \ominus (\dashv X) = \{v \in V : \forall u \ (u \in X \ \text{implies} \ u \ R \ v)\} = \hat{R} \ominus (-X).$$

Proof. In one direction we have $\hat{R} \ominus (-X) \subseteq \hat{R} \ominus (\dashv X)$ since $-X \subseteq \dashv X$.

For the reverse inclusion we give a proof using the special case of the pre-order being a graph so it can be visualized using diagrams as in Section 2.3. Suppose that $v \in \hat{R} \ominus (\dashv X)$, and that $u \in X \cap \dashv X$. There must be some edge, w, which lies in X and is incident with u. As $w \notin \dashv X$ it is impossible that $v \check{R} w$ so $w \ R \ v$ holds. Since R is an $\check{H}$-relation, this implies $u \ R \ v$. This argument can be readily extended to the general case of any preorder.   □

The following result shows that using the pseudocomplement $\neg X$ as the formalization of the outside of the extent X does yield a notion of intent as consisting of those properties holding everywhere in $\neg\neg X$, the expansion of X.

Theorem 10. *If X is an H-set then*

$$\hat{R} \ominus (\neg X) = \{v \in V : \forall u\ (u \in \dashv \neg X \text{ implies } u\,R\,v)\}.$$

Proof. As with Theorem 9 we use the special case of graphs which makes the underlying ideas easily visualized. The case of an arbitrary pre-order is similar. Suppose that $v \in \hat{R} \ominus (\neg X)$ and that $u \in \dashv \neg X$, the expansion of X. If it is not true that $u\,R\,v$ then we have $v\,\hat{R}\,u$ so that $u \in \neg X$, and hence $u \in \neg X \cap \dashv \neg X$. Some edge, say z, must be incident with u and be in $\dashv \neg X$ but not in X. From $\breve{H}\,;R \subseteq R$ we deduce that $z\,R\,v$ cannot hold. But we then have $v\,\hat{R}\,z$ with $z \notin \neg X$ which is a contradiction, so $u\,R\,v$.

Conversely, assume that for every $u \in \dashv \neg X$ we have $u\,R\,v$ for a particular v. If $v \notin (\neg X) \ominus \hat{R}$ there is some $w \notin \neg X$ for which $v\,\hat{R}\,w$. Since $-\neg X \subseteq \dashv \neg X$ we get $w \in \dashv \neg X$ so that $w\,R\,v$ which contradicts $v\,\hat{R}\,w$. $\square$

In the simplest approach to formal concept analysis, an object either has or does not have a particular property. The use of many-valued attributes and scaling is one way to model properties which may hold to some degree. In the case of graphs, hypergraphs and more generally pre-orders, we can distinguish degrees of possession even for single valued attributes.

Definition 10. *Let $x \in U$ and $y \in V$ in the formal context $R : (U, H) \to (V, K)$. Then we say object x **strongly possesses** property y if for all x' where $x\,H\,x'$ or $x'\,H\,x$ we have $x'\,R\,y$.*

In the special case of a graph, a node will posses a property strongly just when it together with any incident edges possess the property. It follows from the requirement that R is an $(\breve{H}, K)$-relation that edges which possess a property must possess it strongly.

Theorem 10 allows us to define a new kind of formal concept.

Definition 11. *Let $R : (U, H) \to (V, K)$ be a formal context. A **strong concept** is a pair (X, Y) where X is an H-set, Y is a K-set, and the following hold:*

$$\dashv \neg X = -(Y \oplus \hat{R})$$
$$Y = \hat{R} \ominus (\neg X).$$

Conceptually the intent, Y, of a strong concept consists of all those properties which hold strongly throughout X. The extent X, of such a concept then consists of those objects which not only have all the properties in Y, but have them strongly. Another view of the intent is that it consists of the properties not merely holding throughout X but holding anywhere 'near' X.

The ability to define both ordinary concepts and strong concepts for a context $R : (U, H) \to (V, K)$ provides a further indication of the expressive power of the algebraic operations from Section 2.3. Clearly this style of investigation can be carried further by considering, for example, the role of the contraction of X in place of the expansion.

4   Application Domains

We discuss how the structure of an $(\check{H}, K)$-relation R can correspond to useful situations. For the first examples it is sufficient to assume that K is just the identity relation on V.

1. Take the edges of a graph to be intervals of time. Every interval has two nodes as its endpoints, but it is also permitted to have nodes (time instants) that are not the endpoint of any interval. This structure gives us an H-set in which we use $u \, H \, u'$ to model the idea that edge u is incident with node u'.

   Now consider the principle:

   > If a property holds throughout a time interval then it must hold at the end points of the interval.

   This principle is just the requirement that $\check{H} \, ; R \subseteq R$.

2. Instead of intervals of time forming a graph, consider edges in a hypergraph that correspond to geographic regions. A region here is an abstract entity that may contain specific point locations, modelled by nodes in the hypergraph. It is permitted to have points that lie in no region. Some (but not all) properties will adhere to this principle:

   > If a geographical region as a whole has a property then any points in the region have the property.

   To take a simple example, the property of being above a certain altitude, or the property of lying within a certain administrative region. In such cases the requirement $\check{H} \, ; R \subseteq R$ formalizes the principle correctly.

3. Granularity provides another example. Suppose we have data about objects at two levels of detail or resolution. These might be obtained by some remote-sensing technology. A high-level object can be represented as a node and a low level one as an edge in a hypergraph. The incidence structure records which low-resolution objects resolve into what sets of high-level ones. Consider the following example in which passage from more detail to less can involve closing up small gaps and ignoring narrow linkages.

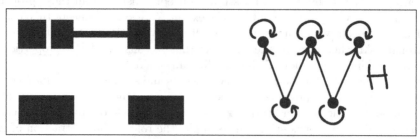

In the diagram the higher resolution version appears at the top. The principle here is

> If low-resolution object α is known to have property v then a higher resolution version of α still has v.

Again, the requirement $\breve{H} \mathbin{;} R \subseteq R$ formalizes the principle correctly.

Considering the case of K not just being the identity on V, the most evident is when the properties are hierarchical, so that $v \; K \; v'$ means that v is more specific than v'. To model the principle:

> if an object has a specific property then it also has any more general version of the property

we need to impose the condition $R \mathbin{;} K \subseteq R$. Of course this in itself can already be handled by existing techniques in formal concept analysis, but the ability to combine it with the three examples just mentioned appears to be a strength of this approach.

5   Conclusions and Further Work

We have combined the extension of mathematical morphology to graphs and hypergraphs with formal concept analysis to establish a foundation for concepts where instead of a set of objects having properties there is a graph or a hypergraph. Although mathematical morphology has been considered in connection with hypergraphs and separately with formal concept analysis before, the study of the interaction between these three models in a single setting is novel. The whole framework depends on the notion of relation between hypergraphs which is a special case of the notion of relation between pre-orders.

It should be possible to build on this foundation in terms of relations and extend some of the work by Priss [Pri06, Pri09] to the relation algebra operations for relations on hypergraphs described by Stell [Ste12]. As the results in Section 3 show, there are a number of potential definitions of formal concept in our generalized setting, and analysing the properties of all of these will take a significant effort.

Another direction is to extend the studies which have investigated connections between rough set theory and formal concept analysis. These studies include [DG03, Yao04, YM09]. The notion of an 'object-oriented concept' [Yao04, secn 2.3] in a context $R \subseteq U \times V$ can readily be expressed in terms of the morphological operations as a pair (X, Y) where $X = Y \oplus \breve{R}$ and $Y = \breve{R} \ominus X$. However, if we have a context with pre-orders (U, H) and (V, K) and R is an $(\breve{H}, K)$-relation, we find that object-oriented concepts provide an H-set X together with a $\breve{K}$-set Y. This contrasts with the situation for Definition 9 where Y is a K-set, so a full account would appear to need to consider the dual subgraphs as well as the subgraphs.

References

[BB11]   Bloch, I., Bretto, A.: Mathematical morphology on hypergraphs: Prelim-
            inary definitions and results. In: Debled-Rennesson, I., Domenjoud, E.,
            Kerautret, B., Even, P. (eds.) DGCI 2011. LNCS, vol. 6607, pp. 429–440.
            Springer, Heidelberg (2011)

[BHR07]   Bloch, I., Heijmans, H.J.A.M., Ronse, C.: Mathematical morphology. In:
            Aiello, M., Pratt-Hartmann, I., van Benthem, J. (eds.) Handbook of Spatial
            Logics, ch. 14, pp. 857–944. Springer (2007)

[BL86]   Bumby, R.T., Latch, D.M.: Categorical constructions in graph theory. Int.
            J. Math. Math. Sci. 9, 1–16 (1986)

[Blo11]   Bloch, I.: Mathematical morphology, lattices, and formal concept analysis.
            In: Napoli, A., Vychodil, V. (eds.) Eighth International Conference on
            Concept Lattices and Their Applications. CEUR Workshop Proceeedings,
            vol. 959 (2011), http://ceur-ws.org/Vol-959/

[BMSW08]   Brown, R., Morris, I., Shrimpton, J., Wensley, C.D.: Graphs of Mor-
            phisms of Graphs. Electronic Journal of Combinatorics 15, (#A1) (2008),
            http://www.combinatorics.org/ojs/

[CL05]   Chen, X., Li, Q.: Formal topology, Chu space and approximable concept.
            In: Belohlavek, R., Snasel, V. (eds.) Proceedings of the CLA 2005 Interna-
            tional Workshop on Concept Lattices and their Applications, pp. 158–165
            (2005), http://ceur-ws.org

[CNS09]   Cousty, J., Najman, L., Serra, J.: Some morphological operators in graph
            spaces. In: Wilkinson, M.H.F., Roerdink, J.B.T.M. (eds.) ISMM 2009.
            LNCS, vol. 5720, pp. 149–160. Springer, Heidelberg (2009)

[DG03]   Düntsch, I., Gediga, G.: Approximation operators in qualitative data anal-
            ysis. In: de Swart, H., Orłowska, E., Schmidt, G., Roubens, M. (eds.)
            TARSKI. LNCS, vol. 2929, pp. 214–230. Springer, Heidelberg (2003)

[DP11]   Dubois, D., Prade, H.: Bridging gaps between several frameworks for the
            idea of granulation. In: IEEE Symposium on Foundations of Computa-
            tional Intelligence (FOCI), pp. 59–65. IEEE Press (2011)

[GW99]   Ganter, B., Wille, R.: Formal Concept Analysis: mathematical foundations.
            Springer (1999)

[HV93]   Heijmans, H., Vincent, L.: Graph Morphology in Image Analysis. In:
            Dougherty, E.R. (ed.) Mathematical Morphology in Image Processing, ch.
            6, pp. 171–203. Marcel Dekker (1993)

[Law86]   Lawvere, F.W.: Introduction. In: Lawvere, F.W., Schanuel, S.H. (eds.) Cat-
            egories in Continuum Physics. Lecture Notes in Mathematics, vol. 1174,
            pp. 1–16. Springer (1986)

[Pri06]   Priss, U.: An FCA interpretation of relation algebra. In: Missaoui, R.,
            Schmidt, J. (eds.) ICFCA 2006. LNCS (LNAI), vol. 3874, pp. 248–263.
            Springer, Heidelberg (2006)

[Pri09]   Priss, U.: Relation algebra operations on formal contexts. In: Rudolph, S.,
            Dau, F., Kuznetsov, S.O. (eds.) ICCS 2009. LNCS, vol. 5662, pp. 257–269.
            Springer, Heidelberg (2009)

[Ros96]   Rosenthal, K.I.: The theory of quantaloids. Pitman Research Notes in
            Mathematics, vol. 348. Longman (1996)

[RZ96]   Reyes, G.E., Zolfaghari, H.: Bi-Heyting algebras, toposes and modalities.
            Journal of Philosophical Logic 25, 25–43 (1996)

[Ser82]   Serra, J.: Image Analysis and Mathematical Morphology. Academic Press, London (1982)

[Ste07]   Stell, J.G.: Relations in Mathematical Morphology with Applications to Graphs and Rough Sets. In: Winter, S., Duckham, M., Kulik, L., Kuipers, B. (eds.) COSIT 2007. LNCS, vol. 4736, pp. 438–454. Springer, Heidelberg (2007)

[Ste10]   Stell, J.G.: Relational granularity for hypergraphs. In: Szczuka, M., Kryszkiewicz, M., Ramanna, S., Jensen, R., Hu, Q. (eds.) RSCTC 2010. LNCS (LNAI), vol. 6086, pp. 267–276. Springer, Heidelberg (2010)

[Ste12]   Stell, J.G.: Relations on hypergraphs. In: Kahl, W., Griffin, T.G. (eds.) RAMiCS 2012. LNCS, vol. 7560, pp. 326–341. Springer, Heidelberg (2012)

[Yao04]   Yao, Y.: A comparative study of formal concept analysis and rough set theory in data analysis. In: Tsumoto, S., Słowiński, R., Komorowski, J., Grzymała-Busse, J.W. (eds.) RSCTC 2004. LNCS (LNAI), vol. 3066, pp. 59–68. Springer, Heidelberg (2004)

[YM09]   Lei, Y., Luo, M.: Rough concept lattices and domains. Annals of Pure and Applied Logic 159, 333–340 (2009)

[ZS06]   Zhang, G.-Q., Shen, G.: Approximable concepts, Chu spaces and information systems. Theory and Application of Categories 17, 80–102 (2006)

Automatic Strengthening
of Graph-Structured Knowledge Bases
Or: How to Identify Inherited Content in Concept Graphs

Vinay Chaudhri, Nikhil Dinesh, Stijn Heymans, and Michael Wessel*

SRI International
Artificial Intelligence Center
333 Ravenswood Avenue, Menlo Park, CA 94025-3493, USA
`firstname.lastname@sri.com`

Abstract. We consider the problem of identifying inherited content in knowledge representation structures called *concept graphs (CGraphs)*. A CGraph is a visual representation of a concept; in the following, CGraphs and concepts are used synonymously. A CGraph is a node- and edge-labeled directed graph. Labeled (binary) edges represent relations between nodes, which are considered instances of the concepts in their node labels. CGraphs are arranged in a taxonomy (is-a hierarchy). The taxonomy is a directed acyclic graph, as multiple inheritance is allowed. A taxonomy and set of CGraphs is called a graph-structured knowledge base (GSKB).

A CGraph can inherit content from other CGraphs – intuitively, if C and D are CGraphs, then C may contain content inherited from D, i.e. labeled nodes and edges "from D" can appear in C, if D is a direct or indirect superconcept of C, or if C contains a node being labeled with either D or some subclass of D. In both cases, C is said to refer to D.

This paper contains three contributions. First, we describe and formalize the problem from a logical point of view and give a first-order semantics for CGraphs. We show that the identification of inherited content in CGraphs depends on some form of hypothetical reasoning and is thus not a purely deductive inference task, as it requires unsound reasoning. Hence, this inference is different from the standard subsumption checking problem, as known from description logics (DLs) [1]. We show that the *provenance problem* (from where does a logical atom in a CGraph get inherited?) strongly depends on the solution to the *co-reference problem* (which existentials in the first-order axiomatization of concepts as formulas denote identical domain individuals?) We demonstrate that the desired inferences can be obtained from a so-called *strengthened GSKB*, which is an augmented variant of the input GSKB. We present an algorithm which augments and strengthens an input GSKB, using model-theoretic notions. Secondly, we are addressing the problem from a graph-theoretic point of view, as this perspective is closer to the actual implementation. We show that we can identify inherited content by computing so-called *concept coverings,* which induce inherited content from superconcepts by means of *graph morphisms*. We argue that the algorithm solves a challenging (NP-hard) problem. Thirdly, we apply the algorithm to the large-scale biological knowledge base from the AURA project [2], and present a preliminary evaluation of its performance.

---

* Corresponding author.

M. Croitoru et al. (Eds.): GKR 2013, LNAI 8323, pp. 180–210, 2014.

1   Introduction

Graph-structured knowledge bases (GSKBs) occur naturally in many application domains, for example, if biological knowledge is modeled graphically by means of *concept graphs (CGraphs)* as in the AURA project [2], see Fig. 1:

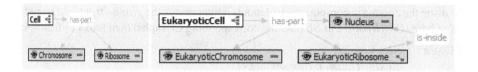

Fig. 1. (Simplified) Concept Graphs for *Cell* and *EukaryoticCell* in AURA

In the AURA project, such CGraphs were modeled by subject-matter experts (SMEs) from the pages of a biology college textbook [3], following a detailed "text to CGraph" encoding process [4]:

S1. Every *Cell* has part a *Ribosome* and a *Chromosome*.
S2. Every *EukaryoticCell* is a *Cell*.
S3. Every *EukaryoticCell* has part a *EukaryoticRibosome*, a *EukaryoticChromosome*, a *Nucleus*, such that the *EukaryoticChromosome* is inside the *Nucleus*.
S4. Every *EukaryoticRibosome* is a *Ribosome*.
S5. Every *EukaryoticChromosome* is a *Chromosome*.

The CGraphs in Fig. 1 naturally represent $S1$ and $S3$. However, the taxonomic information expressed by $S2$, $S4$ and $S5$, is visualized in a separate taxonomy view (hence, the superconcepts are not explicit in the CGraphs in Fig. 1).

Intuitively, a CGraph C *can inherit content* from a CGraph D if D is either a direct or indirect superconcept of C, or if C contains a node with either D or some subconcept of D in its label.

In this paper, we are addressing the following question: **Given a GSKB with a fixed taxonomy, which content in the CGraphs is inherited, and from where?** For example, if we assume that *EukaryoticCell* has *Cell* as a superconcept, as expressed by $S2$, then it also seems plausible to assume that its *EukaryoticChromosome* part was inherited from *Cell* as a *Chromosome* which then got *specialized*. Under this assumption it seems reasonable to assume that the individual represented by the *EukaroticChromosome* node in *EukaryoticCell* is hence identical to the individual represented by the *Chromosome* node in *Cell* – we are saying that those nodes should be *co-referential*. However, this co-reference is not explicit in the CGraphs. We might consider the graphs as *underspecified* as it is neither possible to prove nor to disprove equality of those nodes resp. individuals represented by them in the logical models.

Similar questions of co-referentiality related to inherited content do arise if we use natural language as our primary means of knowledge representation, e.g. consider the sentences $S1$ and $S3$. Is the *EukaryoticChromosome* that $S3$ is talking about actually the *same Chromosome* that is mentioned in $S1$? These kinds of question are studied to some extent in the field of computational linguistics under the term *anaphora resolution* [5], [6]. We will use the term *co-reference resolution* in the following. Obviously, natural language is also underspecified in that sense.

From a *logical point of view*, the sentences $S1$ to $S5$ (and their corresponding CGraphs) naturally correspond to FOPL formulas of the following kind (in the following, we will be using the comma in consequents to denote conjunctions); the *graph-structured knowledge base (GSKB)* of $S1$ to $S5$ (and the corresponding CGraphs) then looks as follows:

S1. $\forall x : Cell(x) \Rightarrow \exists x_1, x_2 :$
$\quad\quad hasPart(x, x_1), Ribosome(x_1),$
$\quad\quad hasPart(x, x_2), Chromosome(x_2)$

S2. $\forall x : EukaryoticCell(x) \Rightarrow Cell(x)$

S3. $\forall x : EukaryoticCell(x) \Rightarrow \exists x_3, x_4, x_5 :$
$\quad\quad hasPart(x, x_3), Euk.Ribosome(x_3),$
$\quad\quad hasPart(x, x_4), Euk.Chromosome(x_4),$
$\quad\quad hasPart(x, x_5), Nucleus(x_5), inside(x_4, x_5)$

S4. $\forall x : Euk.Ribosome(x) \Rightarrow Ribosome(x)$

S5. $\forall x : Euk.Chromosome(x) \Rightarrow Chromosome(x)$

Unary predicates represent *concepts*, and binary predicates predicates *relations*. The concept D is called a *superconcept* of C, with $C \neq D$, if $\forall x : C(x) \Rightarrow D(x), \ldots$ holds. Vice versa, C is called a *subconcept* of D then.

Following our example, we would like to prove that a *EukaryoticCell* has a *Chromosome* part which gets inherited from *Cell*, and indeed, the following entailment holds:

$$\{S1, S2, S4, S5\} \models$$
$$\forall x : EukaryoticCell(x) \Rightarrow \exists y : hasPart(x, y), Chromosome(y),$$

with $S3$ being removed here, as the entailment would hold true trivially otherwise, of course. This testifies that "having a Chromosome part" is inherited from *Cell* to *EukaryoticCell*.

However, a closer look reveals that our question was a bit too general, and that we really ought to be able to prove that *the Chromosome part inside the Nucleus of a EukaryoticCell* should be *the* Chromosome inherited from *Cell*; intuitively, the atoms $hasPart(x, x_4), Chromosome(x_4)$ in $S3$ should be "inherited" from $S1$ (their *provenance* should be *Cell*), and hence, be logically redundant in a sense. In order to check if those atoms from $S3$ are indeed inherited from some other concept in the GSKB to *EukaryoticCell* and hence redundant, we can tentatively remove the atoms under question from $S3$, yielding $S3^-$, and then check if the original $S3$ axiom is entailed. Following this procedure, $S3^-$ reads

S3⁻ $\forall x : EukaryoticCell(x) \Rightarrow \exists x_3, x_4, x_5 :$
$\quad\quad hasPart(x, x_3), Euk.Ribosome(x_3),$
$\quad\quad\quad Euk.Chromosome(x_4),$
$\quad\quad hasPart(x, x_5), Nucleus(x_5), inside(x_4, x_5)$

Note that of the two atoms to be removed from $S3$, namely $hasPart(x, x_4)$, $Chromosome(x_4)$, only the former is actually present in $S3$ – obviously, we do not want to remove $Euk.Chromosome(x_4)$ from $S3$ for this test. Using $S3^-$ in the GSKB instead of $S3$ *unfortunately* yields:

$$\{S1, S2, S3^-, S4, S5\} \not\models S3$$

since already

$$\{S1, S2, S3^-, S4, S5\} \not\models$$
$$\forall x : EukaryoticCell(x) \Rightarrow \exists y_1, y_2 :$$
$$hasPart(x, y_1), Chromosome(y_1), inside(y_1, y_2).$$

The reason that the desired consequence *does not* hold is obviously that there is *no way to prove the co-referentiality / equality* between the *Chromosome* inherited from *Cell*, denoted by x_2 in $S1$, and the *EukaryoticChromosome* denoted by x_4 in $S3^-$ *inside the Nucleus* x_5, and hence, the corresponding model individuals cannot be used for satisfaction / interpretation of y_1, y_2 in the above entailment query. If we could prove that x_2 from $S1$ must be equal to x_4 in $S3^-$ in the context of $EukaryoticCell$, then the entailment would hold. But all we can say is that there is some (potentially different) *Chromosome* part inherited from *Cell* to *EukaryoticCell*, which is not necessarily inside its *Nucleus*. So, we do not get the desired inference *unless* we *strengthen* the axiomatic content of our GSKB somehow such that those equalities hold.

One way of *strengthening* this GSKB in order to *get the desired inferences* is to *skolemize* the existentials, and establish co-references by virtue of equality atoms between ("local and inherited") Skolem function terms, as shown below. We hence call the following GSKB a *strengthened version* of the original GSKB:

S1' $\forall x : Cell(x) \Rightarrow$
$\qquad hasPart(x, f_1(x)), Ribosome(f_1(x)),$
$\qquad hasPart(x, f_2(x)), Chromosome(f_2(x))$

S3' $\forall x : EukaryoticCell(x) \Rightarrow$
$\qquad hasPart(x, f_3(x)), Euk.Ribosome(f_3(x)),$
$\qquad hasPart(x, f_4(x)), Euk.Chromosome(f_4(x)),$
$\qquad hasPart(x, f_5(x)), Nucleus(f_5(x)),$
$\qquad inside(f_4(x), f_5(x))),$
$\qquad f_3(x) = f_1(x), f_4(x) = f_2(x)$

Note that we are now getting the desired inference as follows. First we again remove the atoms under investigation from $S3'$ to get $S3'^-$:

S3'$^-$ $\forall x : EukaryoticCell(x) \Rightarrow$
$\qquad hasPart(x, f_3(x)), Euk.Ribosome(f_3(x)),$
$\qquad\qquad Euk.Chromosome(f_4(x)),$
$\qquad hasPart(x, f_5(x)), Nucleus(f_5(x)),$
$\qquad inside(f_4(x), f_5(x))),$
$\qquad f_3(x) = f_1(x), f_4(x) = f_2(x)$

and then we observe that the following entailment holds:

$$\{S1', S2, S3'^{-}, S4, S5\} \models S3'$$

because for an x satisfying $EukaryoticCell$, we can inherit $hasPart(x, f_2(x))$, $Chromosome(f_2(x))$ from $Cell$, and due to the equality atom $f_4(x) = f_2(x)$, we can establish co-referentiality / equality of the inherited $Chromosome$ with the $Euk.Chromosome$ in $EukaryoticCell$ and hence, the restrictions modeled in $S3'^{-}$ for $f_4(x)$ apply to the inherited $f_2(x)$, hence satisfying all of the necessary conditions in $S3'$. This shows that the $hasPart(x, f_4(x))$, $Chromosome(f_4(x))$ is redundant in $S3'$, or *inherited from somewhere*. More specifically, we can also verify *from where* these atoms are inherited, by removing axioms tentatively from the GSKB one by one, and re-checking the entailment after each removal – for example, if the entailment stops to hold as soon as we remove $S1'$ from the KB, then we know that $S1$ plays a crucial role in entailing these atoms, and that those atoms are, in that sense, "inherited" from $Cell$. Thus, $Cell$ should be their provenance then. This gives us a procedure for computing the provenance of atoms; see below for more details.

We have just demonstrated that inherited content can frequently only be recognized correctly in CGraphs if the GSKB was strengthened, i.e., equality between existentials / Skolem function terms can be proven. Shouldn't it then be *obligatory* that the equalities required for establishing the desired inferences are *explicitly specified in the first place?* We believe that the answer can be *no*. Certainly, the required equalities could be added by hand as in the examples above, but this is often tedious, even if there is some tool support (e.g. a graphical CGraph editor). However, as demonstrated, the input may also be considered *naturally underspecified* in the sense that *co-references are not explicit*. We therefore propose an *automatic co-reference resolution algorithm* which hypothesizes and adds these equality atoms automatically. This algorithm necessarily has to rely on *some sort of guessing,* and is hence in the realm of *hypothetical / logically unsound inference procedures*. The presented algorithm produces a strengthened GSKB in which these co-references are explicit.

Another benefit of a GSKB is that it often *reduces modeling effort.* For example, suppose we updated $Cell$ by saying that its $Ribosome$ is *inside* $Cytosol$:

S1b. $\forall x : Cell(x) \Rightarrow \exists x_1, x_2, x_6 :$
          $hasPart(x, x_1), Ribosome(x_1),$
          $hasPart(x, x_2), Chromosome(x_2),$
          $inside(x_1, x_6), Cytosol(x_6)$

We would like to derive that this also holds for the $Euk.Ribosome(x_3)$ in $EukaryoticCell$ in $S3$ – analog to the case of the $Chromosome$ and $EukraryoticChromosome$ it is reasonable to assume that $Ribosome$ and $EukaryoticRibosome$ are co-referential as well, and indeed, we are getting this inference automatically with

S1b' $\forall x : Cell(x) \Rightarrow$
          $hasPart(x, f_1(x)), Ribosome(f_1(x)),$
          $hasPart(x, f_2(x)), Chromosome(f_2(x)),$
          $inside(f_1(x), f_0(x)), Cytosol(f_0(x))$

as follows:

$$\{S1b', S2, S3', S4, S5\} \models$$
$$\forall x : EukaryoticCell(x) \Rightarrow$$
$$\exists y_1, y_2 : hasPart(x, y_1), Euk.Ribosome(y_1), inside(y_1, y_2), Cytosol(y_2)$$

Note again that this entailment does *not* hold for $\{S1b, S2, S3, S4, S4\}$. the utility of the strengthened GSKB $\{S1b', S2, S3', S4, S4\}$ is hence that we do not need to re-model the fact that the *EukaryoticRibosome* of a *EukaryoticCell* is inside *Cytosol* – we can simply inherit it from *Cell*, and it would be *logically redundant* to add these atoms to $S3'$.

In this paper, we first present the *logical perspective* on this problem. We present a so-called GSKB strengthening algorithm, which, given an input GSKB such as $\{S1b, S2, S3, S4, S5\}$, produces a strengthened GSKB similar to $\{S1b', S2, S3', S4, S5\}$, by using Skolemization and equality atoms between Skolem terms. From the strengthened GSKB we can compute the *provenance of atoms* and *co-references*, hence answering the question *from where did content get inherited?* We will use a model-theoretic notion of *preferred models* in order to characterize the desired inferences that we wish to obtain from the underspecified GSKB, as illustrated by the examples above. The information in the preferred model(s) is then used to produce the strengthened GSKB. Obviously, deciding entailment of atoms, and hence the provenance problem, are in general undecidable in FOPL(=), but decidable in the considered fragment of FOPL(=).

A further contribution of this paper is a description of the *actual implementation, which is best understood and described from a graph-theoretic perspective.* It is obvious that we can define CGraphs as node- and edge labeled graphs. We will argue that the problem of identifying inherited content (the provenance problem) and the co-reference problem can also be understood as solving a variant of the well-known *maximum common subgraph isomorphism (MCS) problem.* This problem, which is NP-hard, can be stated as follows:

Input: Two graphs G1 and G2.
Output: The largest subgraph of G1 isomorphic to a subgraph of G2.

We are considering morphisms instead of isomorphisms, as the node mappings do not have to be injective in our case.[1] Moreover, we also need to be more flexible regarding the node- and edge-label matching conditions, as inherited labels can be specialized in subconcepts. The resulting morphism-based graph algorithm is called the *GSKB covering algorithm* in the following.

We have implemented this novel GSKB covering in a scalable way, and have successfully applied it to the AURA GSKB [2], [4]. The AURA GSKB is the basis of the intelligent question answering textbook Inquire, see [7][2], and currently contains around

---

[1] There is a possibility that two inherited nodes get equated / merged into one node, which means that those two (or more) nodes will be mapped to one node by the morphism – the mapping is obviously no longer injective then.

[2] This video won the AAAI 2012 video award.

6430 concept graphs, with a total of 93,254 edges and 53,322 nodes, hence on average 14.5 edges and 8.2 nodes per concept. The biggest graph has 461 nodes and 1,308 edges. A third contribution of this paper is hence a preliminary evaluation and critical review of the performance of this algorithm on the AURA GSKB.

The paper is structured as follows: We first address the problem from a logical point of view. We formally define CGraphs and the GSKB framework, as well as the required notions of strengthened GSKBs, and the semantic notion of preferred models. We then present the GSKB-strengthening algorithm and show that a strengthened GSKB has models which are preferred models of the original GSKB, hence giving us the desired additional conclusions required to solve the co-reference and provenance problems. We then introduce the graph-based framework underlying the actual implementation. We formally define CGraph morphisms, as well as so-called CGraph patchworks and GSKB coverings, which describe how content is inherited from other CGraphs via morphisms throughout the whole GSKB. We present an algorithm for computing such a GSKB covering, and illustrate that the computed CGraph morphisms can be used to decide provenance and co-referentiality. Next we apply the GSKB covering algorithm to the AURA GSKB and evaluate its performance. We then discuss related work, and conclude with a summary and an outline for future work.

2   Graph Structured Knowledge Bases – The Logical Perspective

As outlined in the Introduction, in this Section we formalize the notion of GSKBs from a first-order logic perspective, and show how provenance of atoms and co-referentiality of variables can be formalized and computed from a strengthened GSKB. We present the so-called GSKB strengthening algorithm.

2.1   Definitions

In the following, we denote an atom or a conjunction of atoms with free variables $\{x, x_1, \ldots, x_n\}$ as $\varphi(x, \boldsymbol{x})$, with $\boldsymbol{x} = (x_1, \ldots, x_n)$. *Graph-structured knowledge bases* (GSKBs) are formulated in first order-logic with equality, FOPL(=). We assume that there is a function $terms$ which returns the terms in a formula, e.g. $terms(R(t_1, t_2)) \triangleq \{t_1, t_2\}$:

Definition 1. *Basic Definitions. Let $\mathcal{C}$ ($\mathcal{R}$) be a countably infinite set of unary (binary) predicate names, and $\mathcal{F} = \{f_1, f_2, \ldots\}$ be a countably infinite set of unary function names – hence, $(\mathcal{C} \cup \mathcal{R}, \mathcal{F})$ is the signature. Elements in $\mathcal{C}$ ($\mathcal{R}$) are called concepts (relations). Moreover, let $\mathcal{X} = \{x, x_1, x_2, \ldots\}$ be a countably infinite set of variables. A GSKB term is a term t such that $t \in \mathcal{X}$, or $t = f_i(x)$, or $t = f_i(f_j(x))$, with $\{f_i, f_j\} \subseteq \mathcal{F}$. Let t, t_1, t_2 be GSKB terms:*

GSKB atoms: *Let $\{C, D\} \subseteq \mathcal{C}$, $R \in \mathcal{R}$, $\{v, w\} \subseteq \mathcal{X}$. Then, $C(v)$ and $C(f_i(x))$ are concept atoms, and $R(v, w)$, $R(x, f_i(x))$ are relation atoms. Moreover, there are equality and inequality atoms of the following form: $f_i(x) = f_j(x)$, $f_i(x) = f_j(f_k(x))$, $f_j(f_k(x)) = f_i(x)$, and $f_i(x) \neq f_j(x)$, with i, j, k pairwise unequal.*

GSKB rule: *For a concept C, a formula $\rho_C \triangleq \forall x : C(x) \Rightarrow \exists! \boldsymbol{x} : \varphi(x, \boldsymbol{x})$ is called a GSKB rule, where $\varphi(x, \boldsymbol{x}) = \bigwedge_{i \in 1 \ldots m} \alpha_i$ is finite conjunction of GSKB atoms. This is shorthand for $\forall x : C(x) \Rightarrow \exists \boldsymbol{x} : pairwise\_unequal(x, \boldsymbol{x}) \wedge \varphi(x, \boldsymbol{x})$, $\boldsymbol{x} = (x_1, \ldots, x_n)$, with the macro $pairwise\_unequal(x, \boldsymbol{x}) \triangleq \bigwedge_{1 \le i < j \le n} x_i \neq x_j \wedge \bigwedge_{1 \le i \le n} x_i \neq x$.*
For a concept C with $\rho_C = \forall x : C(x) \Rightarrow \exists! \boldsymbol{x} : \varphi(x, \boldsymbol{x})$, denote $\varphi(x, \boldsymbol{x}) = \bigwedge_{i \in 1 \ldots m} \alpha_i$ as a set by $\tau_C \triangleq \{\alpha_1, \ldots, \alpha_m\}$, and $terms(C) \triangleq \bigcup_{\alpha \in \tau_C} terms(\alpha)$. We require that the terms in $terms(\rho_C)$ are connected to x: for all $t \in terms(C)$, $connected(x, t)$ holds, where connected is defined as follows: $connected(t_1, t_2)$ holds if $\{R(t_1, t_2), R(t_2, t_1)\} \cap \tau_C \neq \emptyset$, or there is some t s.t. $connected(t_1, t)$ and $connected(t, t_2)$ holds.

GSKB: *A finite set of GSKB rules Σ in which there is at most one rule per concept.*

Input GSKB: *A GSKB which is function-free and without equality atoms.*

Auxiliary notions: *Given a GSKB Σ, we refer to the set of concepts used in Σ as $concepts(\Sigma)$, and $\tau_{C,\Sigma}$ to refer to the consequent of $\rho_C \in \Sigma$. We extend the other definitions to accept a Σ argument as well, e.g., $terms(C, \Sigma)$, etc.*

For example, $\{S1b, S2, S3, S4, S5\}$ is an (underspecified) input GSKB, and $\{S1b', S2, S3', S4, S5\}$ is a *strengthened* (output) GSKB; however, we need to replace the $\exists$ quantifier with $\exists!$. The formal definition of *strengthened* GSKB is given below. Note that sometimes the strengthening algorithm will not add anything, and hence output will equal input, e.g. for $\{S4, S5\}$.

We require that an input GSKB must be *coherent*:

Definition 2. *Coherent GSKB and coherent model. A GSKB Σ is coherent if there is standard first-order model $\mathcal{I} = (\Delta_{\mathcal{I}}, \cdot^{\mathcal{I}})$, $\mathcal{I} \models \Sigma$, in which every concept C mentioned in Σ is interpreted in a non-empty way: $C^{\mathcal{I}} \neq \emptyset$. Such a model is called a* coherent *model.*

Moreover, we define standard notions such as *superconcepts* as follows:

Definition 3. *Auxiliary Definitions. Let C be a concept, Σ be a GSKB. We then define the following functions and predicates w.r.t. Σ:*

- $asserted\_types(C, \Sigma) \triangleq \{D \mid D(t) \in \tau_{C,\Sigma}\}$
- $has\_asserted\_type_\Sigma(C, D)$
    iff $D \in asserted\_types(C, \Sigma)$
- $asserted\_superconcepts(C, \Sigma) \triangleq \{D \mid D(x) \in \tau_{C,\Sigma}\}$
- $has\_asserted\_superconcept_\Sigma(C, D)$
    iff $D \in asserted\_superconcepts(C, \Sigma)$
- $superconcepts(D, \Sigma)$
    $\triangleq \{E \mid has\_asserted\_superconcept^+_\Sigma(D, E)\}$
- $has\_superconcept_\Sigma(C, D)$
    iff $D \in superconcepts(C, \Sigma)$
- $all\_types_\Sigma(C)$
    $\triangleq \{E \mid D \in asserted\_types(C, \Sigma),$
            $E \in superconcepts(D, \Sigma)\}$
        $\cup \; superconcepts(C, \Sigma)$

– $has\_type_\Sigma(C, D)$ iff $D \in all\_types(C, \Sigma)$

where R^+ denotes the transitive closure of relation R.

We require that the relations $has\_superconcept_\Sigma$ and has_type_Σ are irreflexive and define:

Definition 4. *Admissible GSKB. An input GSKB Σ is called admissible if Σ is coherent, $has\_superconcept_\Sigma$ and has_type_Σ are irreflexive, and if there are no implied concept atoms in the rules: for all $C \in concepts(\Sigma)$, if $D(t) \in \tau_{C,\Sigma}$, then for all $E \in superconcepts(D, \Sigma)$: $E(t) \notin \tau_{C,\Sigma}$.*

The following is straightforward:

Proposition 1. *Every admissible GSKB Σ has a coherent, finite model.*

Proof. Given that we do not support negation of concepts or relations, and given that inequality atoms are only introduced by the $\exists!$ quantor, inconsistencies such as $x \neq x$ cannot occur. Moreover, since GSKB $has\_superconcept_\Sigma$ and has_type_Σ are irreflexive, the GSKB is acyclic, and the consequent of every rule can be "unfolded", analog to the unfolding of an acyclic TBox in description logics [1]. This produces a finite consequent for every rule. Next, for every $\rho_C \in \Sigma$, C can be instantiated s.t. $i_C \in C^\mathcal{I}$ holds, and we can easily satisfy the existentials in the consequent by creating one instance per variable. The process terminates and produces a model of Σ which is coherent and finite.

We need a notion of connectedness on models:

Definition 5. *Predicate connected on models. Let $\mathcal{I} = (\Delta_\mathcal{I}, \cdot^\mathcal{I})$ be a model of Σ. For $i, j \in \Delta_\mathcal{I}$, we define $connected_\mathcal{I}(i, j)$ if, for some $R \in \mathcal{R}$, $\{(i, j), (j, i)\} \cap R^\mathcal{I} \neq \emptyset$, or there is some $k \in \Delta_\mathcal{I}$ s.t. $connected_\mathcal{I}(i, k)$ and $connected_\mathcal{I}(k, j)$.*

In the following we are considering admissible GSKBs only, and we are interested in their *preferred models*. The intuition behind the notion of a preferred model is the following: for every concept C, there should be a *prototypical model* of C which is *not* also a model of some non-superconcept of C, in the form of a connected graph that "mirrors" the atoms in $\tau_{C,\Sigma}$ – due to the *pairwise_unequal* macro there will be at least one individual per variable in ρ_C in this model. Moreover, the prototypical model for C also contains inherited "graphs" from concepts in $all\_types_\Sigma(C)$. Hence, the graph satisfying the atoms $\tau_{C,\Sigma}$ is only a subgraph of the full model for C. Most importantly, the notion of a preferred model captures the intuition that inherited content can be specialized, and hence should give rise to co-references: in the prototypical model for $EukaryoticCell$, the $Chromosome$ inherited from its superclass $Cell$ will be represented by the same individual as its own local $Euk.Chromosome$. Note that this minimizes the extension of $Chromosome$. The same argument applies to arbitrary conjunctions: we will *not* identify the inherited $Chromosome$ with the local $Euk.Ribosome$, as this would result in a model in which $Chromosome \wedge Euk.Ribosome$ is interpreted non-empty, and there are models in which this conjunction is interpreted by the empty set. These intuitions are formalized as follows:

Definition 6. *Preferred model of admissible GSKB Σ. Let Σ be an admissible GSKB, and $\mathcal{I} \models \Sigma$ be a coherent finite model. Then, $\mathcal{I}$ is called* preferred *if the following holds:*

1. *for every concept $C \in concepts(\Sigma)$, there is (at least) one $i \in C^{\mathcal{I}}$ s.t. for all D, if $has\_superconcept(D, C)$, then $i \notin D^{\mathcal{I}}$ – hence, there is at least one element which is "unique" to C, and denoted by i_C.*

2. *for $C \in concepts(\Sigma)$, define $participants_{\mathcal{I}}(C) \triangleq \{j \mid connected_{\mathcal{I}}(i_C, j)\}$. Then, for all $C, D \in concepts(\Sigma)$, with $C \neq D$, the following holds: $participants_{\mathcal{I}}(C) \cap participants_{\mathcal{I}}(D) = \emptyset$.*

3. *for every non-empty subset $\mathcal{CS} \subseteq concepts(\Sigma)$, there is no preferred model $\mathcal{I} \neq \mathcal{I}'$, with $\Delta_{\mathcal{I}'} \subseteq \Delta_{\mathcal{I}}$ s.t. $\bigcap_{C \in \mathcal{CS}} C^{\mathcal{I}'} \subset \bigcap_{C \in \mathcal{CS}} C^{\mathcal{I}}$.*

Consider the preferred models of $\{S1, S2, S3, S4, S5\}$. We are forced to have at least one "unique" *Cell* which is not a *EukaryoticCell*, due to 1. Otherwise, every *Cell* would acquire the properties of *EukaryoticCell*s, which is not desirable. Moreover, none of the individuals connected to that unique *Cell* are shared by another concept, due to 2. Hence, the concept models have the forms of "non-overlapping graphs", and inherited content is "mapped in". We are forced to minimize the extension of every concept, as well as of every conjunction of concepts. This prevents models in which, for example, $Ribosome^{\mathcal{I}} \cap Euk.Chromosome^{\mathcal{I}} \neq \emptyset$ holds, as there are smaller models in which they are interpreted disjointly: $Ribosome^{\mathcal{I}} \cap Euk.Chromosome^{\mathcal{I}} = \emptyset$. Note that the inequality atoms in Σ only prevent "merging" of variables within the same formula, but the individual for $Chromosome(x_2)$ inherited from *Cell* could in principle be made co-referential with the local $Euk.Ribosome(x_3)$ in *EukaryoticCell*. *This is prevented in a preferred model.* Also, looking at the model of *EukaryoticCell*, the co-reference between the from *Cell* inherited $Chromosome(x_2)$ and its own local $Euk.Chromosome(x_4)$ is made explicit, since this will result in the smallest (extension of) $Chromosome^{\mathcal{I}}$. A model in which a *EukaryoticCell* would have two different *Chromosome*s would be larger and in violation to 3. So, we only make those conjunction true in a preferred model that we *have to make true* - for example, $Cell^{\mathcal{I}} \cap EukaryoticCell^{\mathcal{I}} \neq \emptyset$, due to $S2$, and there is no model in which this conjunction is interpreted by a smaller set.

Note that a preferred model is not a "minimal" model in the classical sense. Consider $\forall x : C(x) \Rightarrow \exists! x1 : R(x, x1), D(x1), \forall x : SubC(x) \Rightarrow \exists! x2 : C(x), R(x, x2), E(x2)$. In the classical minimal model $\mathcal{I}$, we would have $\#\Delta_{\mathcal{I}} = 2$, and it would satisfy $D \wedge E$. Also, $C^{\mathcal{I}} = SubC^{\mathcal{I}}$. But this is not what we want. It violates 1, 2, as well as 3. The preferred model will need at least 5 nodes.

In principle, there can be more than one preferred model and hence, more than one strengthened version of the GSKB. For example, consider the GSKB

$$C(x) \Rightarrow \exists! x_1 : R(x, x_1), E(x_1)$$
$$SubC(x) \Rightarrow \exists! x_2, x_3 : C(x),$$
$$R(x, x_2), E(x_2), F(x_2),$$
$$R(x, x_3), E(x_3), G(x_3).$$

Here, x_1 in C can be co-referential with either x_2 in $SubC$, or with x_3.

In the next section, we will show the following constructively, by specifying an algorithm which constructs a preferred model for a given admissible GSKB Σ:

Proposition 2. *Every admissible GSKB has a preferred model.*

We can now state the purpose of the GSKB strengthening algorithm more clearly. Given an admissible GSKB Σ (note that this is an input GSKB), we are interested in finding a strengthened version of Σ:

Definition 7. *Strengthened version of Σ. Given an admissible (input) GSKB Σ, we are calling Σ' a strengthened version of Σ if the following holds:*

1. *for every rule $\rho_C \in \Sigma$, there is a rule $\rho'_C \in \Sigma'$ that uses only the variable x: $terms(\rho'_C) \cap \mathcal{X} = \{x\}$.*
2. *if $\mathcal{I}' \models \Sigma'$ is a standard first-order model of Σ' which is coherent, then $\mathcal{I}' \models \Sigma$, and $\mathcal{I}'$ is a preferred model for Σ. Hence, $\Sigma' \models \Sigma$.*

From a strengthened GSKB, we can decide provenance and co-reference as follows:

Definition 8. *Provenance and co-reference determination. Let C be a concept, Σ' be a strengthened GSKB, and $\mathcal{P} \subseteq \tau_{C,\Sigma'}$. With $\beta = \bigwedge_{\alpha \in \mathcal{P}} \alpha$, we then say that β (and hence all the atoms in $\mathcal{P}$) are*

- *local (or asserted) in C if*
  $$\Sigma' \setminus \{\rho_C\} \cup \{\forall x : C(x) \Rightarrow \bigwedge_{\alpha \in \tau_{C,\Sigma'} \setminus \mathcal{P}} \alpha\}$$
  $$\not\models \forall x : C(x) \Rightarrow \beta,$$
- *and inherited otherwise. More specifically, β (and $\mathcal{P}$) is inherited from D, iff $D(t) \in \tau_{C,\Sigma'}$, and $\beta' = \bigwedge_{\alpha \in \mathcal{P}'} \alpha$ with $\mathcal{P}' = \{\alpha_{[f_i(t) \Rightarrow f_i(x)]} \mid \alpha \in \mathcal{P}\}$ is local in D, and there is no more general $SupD$ with $has\_superconcept_{\Sigma'}(D, SupD)$ such that β (and $\mathcal{P}$) is inherited from $SupD$.*

Moreover, given concepts C, D, two GSKB terms $t_1 \in terms(C)$, $t_2 \in terms(D)$ are said to be co-referential in Σ' iff either $t_1 = t_2 = x$, $t_1 = f_i(x)$, $t_2 = f_j(f_k(x))$, or $t_2 = f_i(x)$, $t_1 = f_j(f_k(x))$, and $\Sigma' \models (\forall x : C(x) \Rightarrow f_i(x) = f_j(x)) \vee (\forall x : D(x) \Rightarrow f_i(x) = f_j(x))$.

Note that a conjunction β is local as soon as *some* atom is already local. Hence, if a complex conjunction β (resp. $\mathcal{P}$) is local, this does *not* mean that *all* its atoms have to be local – some atoms may be inherited.

Proposition 3. *Provenance and co-reference are decidable in a strengthened GSKB Σ'.*

The proof is given in the next Section.

2.2 Constructing a Strengthened GSKB

The algorithm works by performing the following steps:

1. Produce the skolemized version of Σ, Σ_S, by bringing every rule in Σ into Skolem normal form. The skolemized axioms contain no nested function terms, only terms of the form $f_i(x)$ and x. Let $\mathcal{O} \triangleq \{o_C \mid C \in concepts(\Sigma)\}$ be a set of constants, and also add $\{C(o_C) \mid C \in concepts(\Sigma)\}$ to Σ_S.

2. Construct the *minimal Herbrand model* $\mathcal{I}_{\mathcal{H}} = (\Delta_{\mathcal{H}}, \cdot^{\mathcal{I}_{\mathcal{H}}})$ of Σ_S. The minimal Herbrand model is unique and finite, given that Σ is admissible (and does not contain disjunctions in the consequents). Note that the minimal Herbrand model will automatically satisfy the inequality atoms, and it will also satisfy points 1 and 2 from Definition 6, due to the set of constants $\mathcal{O}$ which are instantiated as $\{C(o_C) \mid C \in concepts(\Sigma)\} \subseteq \Sigma_S$, and with the exception of x, there are no shared terms in the rules of Σ_S, as Skolemization creates fresh function symbols for every variable. Thus, o_C represents the root individual of the unique model for concept C, with $o_C^{\mathcal{I}_{\mathcal{H}}} = i_C, i_C \in C^{\mathcal{I}_{\mathcal{H}}}$.
3. Transform $\mathcal{I}_{\mathcal{H}}$ into a preferred model of Σ, $\mathcal{I}_{\mathcal{A}} = (\Delta_{\mathcal{A}}, \cdot^{\mathcal{I}_{\mathcal{A}}})$. $\Delta_{\mathcal{A}}$ is the quotient set of $\Delta_{\mathcal{H}}$ under the $=$ equivalence relation, $\Delta_{\mathcal{A}} = \Delta_{\mathcal{H}} \backslash =$. Hence, the elements of $\Delta_{\mathcal{A}}$ represent the equivalence classes of equated Skolem ground terms from the Herbrand universe $\Delta_{\mathcal{H}}$. This step is non-deterministic, as there may be more than one preferred model for Σ.
4. Use $\mathcal{I}_{\mathcal{A}}$ to construct a strengthened GSKB Σ' from Σ_S which is satisfied by that model. Use the equivalent clusters in $\Delta_{\mathcal{A}}$ to generate equality atoms.
5. From Σ' it is possible to decide the provenance and the co-reference problem, on a syntactic basis.

Since steps 1 and 2 are standard and well-know [8], let us define the algorithm for step 3. We need two more utility notions before we can proceed:

Definition 9. *Relations $\mathcal{E}$ and $\mathcal{U}$, and equivalence classes. Let $\mathcal{I}_{\mathcal{H}} = (\Delta_{\mathcal{H}}, \cdot^{\mathcal{I}_{\mathcal{H}}})$ be the minimal unique Herbrand model after step 2 of Σ_S above. Let $\mathcal{E}$ be a binary relation over terms from the Herbrand universe $\Delta_{\mathcal{H}}$, and define*

$$closure(\mathcal{E}) \triangleq \bigcup_{C \in concepts(\Sigma), k \in \Delta_{\mathcal{H}}}$$
$$\{(f_1(k), f_2(k)) \mid (f_1(o_C), f_2(o_C)) \in \mathcal{E}^{\circledast}\} \cup$$
$$\{(f_1(f_2(k)), f_3(k)) \mid (f_1(f_2(o_C)), f_3(o_C)) \in \mathcal{E}^{\circledast}\} \cup$$
$$\{(f_1(f_2(k)), f_3(f_4(k))) \mid (f_1(f_2(o_C)), f_3(f_4(o_C))) \in \mathcal{E}^{\circledast}\}$$

where $\cdot^{\circledast}$ denotes the reflexive, symmetric, and transitive closure *of a relation. Let $[i]^{\mathcal{E}} \triangleq \{j \mid (i,j) \in closure(\mathcal{E})\}$. Moreover, let $\mathcal{U} \triangleq \{[i]^{\mathcal{E}} \neq [j]^{\mathcal{E}} \mid i_1 \in [i]^{\mathcal{E}}, j_1 \in [j]^{\mathcal{E}}, C \in concepts(\Sigma), (i_1 \neq j_1) \in \tau_{C,\Sigma_s}\}$ be the set of inequality atoms.*

Intuitively, $(i,j) \in \mathcal{E}$ represents $i = j$, and $[i]^{\mathcal{E}}$ represents the equivalence class of i. The relation $\mathcal{E}$ (and hence the equivalence classes) will grow as pairs of equated individuals / terms are added by the algorithm given below. Intuitively, the closure operator makes sure that whenever two terms starting from the same root node o_C are equated in the unique model of C, that then this equality will also hold for all its C instantiations in other parts of the model. Note that also $\mathcal{U}$ will grow, representing inferences such as $i \neq j, k \neq l, j = k \Rightarrow i \neq l$.

The algorithm can now be stated as follows:

Algorithm 1. *Construction of a preferred model for Σ. Let $\mathcal{I}_{\mathcal{H}} = (\Delta_{\mathcal{H}}, \cdot^{\mathcal{I}_{\mathcal{H}}})$ be the minimal unique Herbrand model of Σ_S after step 2 above.*

1. *define* $hasRoot(i) \triangleq o_C$ *iff* $connected_{\mathcal{I_H}}(o_C, i)$ *holds, for every* $C \in concepts(\Sigma)$.
2. *then, non-deterministically apply the following* merging rule *on the model as long as it is applicable:*

  if *there are individuals* $i, j \in \Delta_{\mathcal{H}}$, $i \neq j$, *with* $hasRoot(i) = hasRoot(j) = o_C$ *and* $ind\_types(i) \subseteq ind\_types(j)$, $i \notin [j]^{\mathcal{E}}$, $[i]^{\mathcal{E}} \neq [j]^{\mathcal{E}} \notin \mathcal{U}$, **then**
  $\mathcal{E} \triangleq \mathcal{E} \cup \{(i, j)\}$.

  Assume the rule application stops with a global maximum of inequality atoms s.t. $\#\mathcal{U}$ *is maximized. Since this is a non-deterministic algorithm, such a run exists, and we can assume that the non-deterministic algorithm will produce it.*

3. *define* $\mathcal{I_A} = (\Delta_{\mathcal{A}}, \cdot^{\mathcal{I_A}})$ *as follows:*

  $\Delta_{\mathcal{A}} \triangleq \{[i]^{\mathcal{E}} \mid i \in \Delta_{\mathcal{H}}\}$, *and for all* $C \in concepts(\Sigma) : C^{\mathcal{I_A}} \triangleq \{[i]^{\mathcal{E}} \mid i \in C^{\mathcal{I_H}}\}$, *for all* $R \in \mathcal{R} : R^{\mathcal{I_A}} \triangleq \{([i]^{\mathcal{E}}, [j]^{\mathcal{E}}) \mid (i, j) \in R^{\mathcal{I_H}}\}$.

The algorithm terminates, since $\mathcal{I_H}$ is finite, so there is a finite number of merging possibilities in the rule. The solution which maximizes $\#\mathcal{U}$ can obviously be found by search in a deterministic version.

Lemma 1. $\mathcal{I_A} = (\Delta_{\mathcal{A}}, \cdot^{\mathcal{I_A}})$ *is a preferred model for* Σ.

Proof. Obviously, $\mathcal{I_A}$ is finite and coherent, as it was constructed by the algorithm based on the unique finite Herbrand model. Assume that $\mathcal{I_A}$ is not a preferred model for Σ. By construction, $\mathcal{I_A}$ is a model of Σ_S, as the merging rule preserves the model character of $\mathcal{I_H}$. Since $\mathcal{I_H}$ is a model of the skolemized version, it is also a model of Σ, since $\Sigma_S \models \Sigma$ for the skolemized GSKB [8]. Hence, $\mathcal{I_A}$ is a model of Σ, also.

It remains to show that it is preferred. Assume that it is not. Since points 1 and 2 from Definition 6 are already satisfied by construction, only 3 can be violated. Then, there must be some other model $\mathcal{I}'$ and some $\mathcal{CS} \subseteq concepts(\Sigma)$ such that $\bigcap_{C \in \mathcal{CS}} C^{\mathcal{I}'} \subset \bigcap_{C \in \mathcal{CS}} C^{\mathcal{I}}$, witnessed by $[i]^{\mathcal{E}} \in \bigcap_{C \in \mathcal{CS}} C^{\mathcal{I}}$ with $[i]^{\mathcal{E}} \notin \bigcap_{C \in \mathcal{CS}} C^{\mathcal{I}'}$.

1. If $\bigcap_{C \in \mathcal{CS}} C^{\mathcal{I}'} = \emptyset$, then this violates the assumption that $\Sigma_{\mathcal{H}}$ was a minimal Herbrand model (which does not make things true without need). Hence, $\bigcap_{C \in \mathcal{CS}} C^{\mathcal{I}} = \emptyset$ as well, which contradicts $[i]^{\mathcal{E}} \in \bigcap_{C \in \mathcal{CS}} C^{\mathcal{I}}$.
2. Assume $\mathcal{CS} = \{D\}$ is a single concept name. As $\mathcal{I_H}$ was a minimal model, the existence of i, with $i \in [i]^{\mathcal{E}}$, is somehow enforced by Σ_S, hence there is some term $t_i \in terms(C, \Sigma_S)$ with $D \in ind\_types(t_i)$. Moreover, for the same reason, $D^{\mathcal{I}'} \neq \emptyset$, as otherwise it wouldn't be a model, but $i \notin D^{\mathcal{I}'}$. Consequently, there is some $j \in D^{\mathcal{I}'}$ with $i \neq j$. Then, there must also be some $t_j \in terms(C, \Sigma_S)$ with $D \in ind\_types(t_j)$, with $t_i \neq t_j$.
There are a couple of cases:

  (a) Assume $ind\_type(t_i) \subseteq ind\_types(t_j)$
   i. if $C' = C$ and hence $hasRoot(i) = hasRoot(j) = C$, then $(t_i \neq t_j) \notin \tau_{C, \Sigma_S}$ and $[i]^{\mathcal{E}} \neq [j]^{\mathcal{E}} \notin \mathcal{U}$, as otherwise $\mathcal{I}$ would not be a model. But then, the merging rule would have been applied and merged i and j, such that $[i]^{\mathcal{E}} = [j]^{\mathcal{E}} = \{i, j\}$. Rule application could not have been blocked by the precondition $[i]^{\mathcal{E}} \neq [j]^{\mathcal{E}} \notin \mathcal{U}$, because $\mathcal{I_A}$ was produced by a run in which $\#\mathcal{U}$ was maximized. This means that the rule will be applicable and equate i and j, contradicting the assumption that the algorithm has terminated.

     ii. otherwise, $C \neq C'$, then we don't have to worry: as stated in Definition 6, $participants_{\mathcal{I}_A}(C) \cap participants_{\mathcal{I}_A}(C') = \emptyset$.

(b) Assume $ind\_type(t_j) \subseteq ind\_types(t_i)$: analog to the previous case.

(c) Assume $ind\_type(t_i) \not\subseteq ind\_types(t_j)$. Then there is some $E \in ind\_type(t_i), E \notin ind\_types(t_j)$. As $\mathcal{I}_A$ was a minimal Herbrand model, and there is no way for $[i]^{\mathcal{E}}$ to "vanish" from $E^{\mathcal{I}_A}$, there must be $[i]^{\mathcal{E}} \in E^{\mathcal{I}_A}$ and hence $[i]^{\mathcal{E}} \in \bigcap_{C \in \mathcal{CS}} C^{\mathcal{I}'}$. Contradiction.

3. If $\mathcal{CS} = \{D_1, \ldots, D_n\}$, then there must already be some $\mathcal{CS}' = \{D_m, D_n\}, \mathcal{CS}' \subseteq \mathcal{CS}$ for which we have such an i. If $has\_superconcept(D_m, D_n)$ or vice versa, then there is already some $\mathcal{CS}' = \{D_m\}$, and this is handled by 2. Otherwise, D_m, D_n are not in a sub/superconcept relationship, and corresponding instances are not getting merged by the merging rule. But similar to 2c), this will lead us to conclude that $[i]^{\mathcal{E}} \in \bigcap_{C \in \mathcal{CS}} C^{\mathcal{I}'}$, contradicting the assumption.

Hence, $\mathcal{I}_A$ is a preferred model. Note that this proves Proposition 2.

For what remains to be shown is how we can compute a strengthened GSKB from Σ_S and $\mathcal{I}_A$.

Definition 10. *Construction of strengthened GSKB Σ'. Let Σ_S be the skolemized version of the admissible GSKB, and $\mathcal{I}_A$ be a preferred model of Σ. We then rewrite the rules in Σ_S as follows; note that $\alpha_{[t_1 \Rightarrow t_2]}$ means "in α, substitute all occurrences of t_1 with t_2":*

$\Sigma' \triangleq \{rewrite(\rho_C, terms(C, \Sigma_S)) \mid \rho_C \in \Sigma_S\}, with$

$rewrite(\rho_C, terms) \triangleq C(x) \Rightarrow$

$\bigwedge_{\alpha \in \tau_C, \Sigma_S} \alpha \wedge$

$\bigwedge_{t \in terms, t \neq o_C} hasRoot(t, x)_{[o_C \rightarrow x]} \wedge$

$\bigwedge_{t_1, t_2 \in terms, t_1 \neq t_2} t_1 \neq t_{2[o_C \rightarrow x]} \wedge$

$\bigwedge_{t_1 \in terms, t_2 \in [t_1]} t_1 = t_{2[o_C \rightarrow x]}$

In addition, we need the following axioms:

1. $\Sigma' \triangleq \Sigma' \cup \{C(o_C) \mid C \in concepts(\Sigma)\}$
2. $\Sigma' \triangleq \Sigma' \cup \{o_C \neq o_D \mid C, D \in concepts(\Sigma), C \neq D\}$
3. $\Sigma' \triangleq \Sigma' \cup \{\forall x, y, z :$
   $hasRoot(x, y), hasRoot(y, z) \Rightarrow hasRoot(x, z)\}$
4. $\Sigma' \triangleq \Sigma' \cup \{\forall x, y :$
   $hasRoot(x, o_C), hasRoot(y, o_D) \Rightarrow x \neq y\},$
   for all $C, D \in concepts(\Sigma), C \neq D$.

Lemma 2. *If $\mathcal{I} \models \Sigma'$, then $\mathcal{I}$ is a preferred model for Σ.*

Proof. As Σ' has been constructed from Σ_S by adding equality atoms to explicitly represent the co-references with inherited Skolem function successors, which have been identified by the merging rule from a preferred model of Σ, it is clear that any model of Σ' will force the same co-references, and hence, satisfy point 3 in Definition 6. Moreover, point 1 in Definition 10 makes sure that we have non-empty concept models for every concept by requiring an instance, hence satisfying condition 1 in Definition 6. Point 2 in Definition 10 enforces distinctness between those constants, and point 3 declares *hasRoot* as a transitively closed relation. In combination with the added *hasRoot* atoms in Σ', and with the axioms in point 4 of Definition 10, this ensures that condition 2 in Definition 6 is satisfied, requiring that the unique concept models do not overlap (no sharing of participants).

Let us return to our example. For $\Sigma = \{S1b, S2, S3, S4, S5\}$ we will get Σ_S as follows:

$$Cell(x) \Rightarrow$$
$$hasPart(x, f_1(x)), Ribosome(f_1(x)),$$
$$hasPart(x, f_2(x)), Chromosome(f_2(x)),$$
$$inside(f_1(x), f_0(x)), Cytosol(f_0(x)),$$
$$pairwise\_unequal(\{x, f_0(x), f_1(x), f_2(x)\})$$
$$EukaryoticCell(x) \Rightarrow Cell(x)$$
$$hasPart(x, f_3(x)), Euk.Ribosome(f_3(x)),$$
$$hasPart(x, f_4(x)), Euk.Chromosome(f_4(x)),$$
$$hasPart(x, f_5(x)), Nucleus(f_5(x)),$$
$$inside(f_4(x), f_5(x)),$$
$$pairwise\_unequal(\{x, f_3(x), f_4(x), f_5(x)\})$$
$$Cell(o_{Cell}), Euk.Cell(o_{Euk.Cell}), Ribosome(o_{Ribosome}) \ldots$$

If we look at the minimal Herbrand model of Σ_S, we find that the following atoms are satisfied for $o_{Euk.Cell}$:

$$hasPart(o_{Euk.Cell}, f_1(o_{Euk.Cell})),$$
$$hasPart(o_{Euk.Cell}, f_2(o_{Euk.Cell})),$$
$$inside(f_1(o_{Euk.Cell}), f_0(o_{Euk.Cell})),$$
$$Ribosome(f_1(o_{Euk.Cell})),$$
$$Chromosome(f_2(o_{Euk.Cell})),$$
$$Cytosol(f_0(o_{Euk.Cell})),$$
$$hasPart(o_{Euk.Cell}, f_3(o_{Euk.Cell})),$$
$$hasPart(o_{Euk.Cell}, f_4(o_{Euk.Cell})),$$
$$hasPart(o_{Euk.Cell}, f_5(o_{Euk.Cell})),$$
$$inside(f_4(o_{Euk.Cell}), f_5(o_{Euk.Cell})),$$
$$Euk.Ribosome(f_3(o_{Euk.Cell})),$$
$$Euk.Chromosome(f_4(o_{Euk.Cell})),$$
$$Nucleus(f_5(o_{Euk.Cell})),$$

Moreover, there are pairwise inequality atoms between $o_{Euk.Cell}, f_3(o_{Euk.Cell})$, $f_4(o_{Euk.Cell}), f_5(o_{Euk.Cell})$ and between $o_{Euk.Cell}, f_0(o_{Euk.Cell}), f_1(o_{Euk.Cell})$, $f_2(o_{Euk.Cell})$.

If we next look at $\mathcal{I}_A$, we will find that $[f_3(o_{Euk.Cell})) = [f_1(o_{Euk.Cell})] = \{f_3(o_{Euk.Cell}), f_1(o_{Euk.Cell})\}$ holds, and likewise $[f_4(o_{Euk.Cell})) = [f_2(o_{Euk.Cell})] = \{f_2(o_{Euk.Cell}), f_4(o_{Euk.Cell})\}$. Hence, the desired co-references have been established, e.g., the from *Cell* inherited $Ribosome(f_1(o_{Euk.Cell}))$ is identified as being co-referential with the "local" $Euk.Ribosome(f_3(o_{Euk.Cell}))$.

The abridged strengthened GSKB Σ' then looks as follows:

$Cell(x) \Rightarrow$
  $hasPart(x, f_1(x)), Ribosome(f_1(x)),$
  $hasPart(x, f_2(x)), Chromosome(f_2(x)),$
  $hasRoot(f_1(x), x), hasRoot(f_2(x), x),$
  $pairwise\_unequal(\{x, f_1(x), f_2(x)\})$
$EukaryoticCell(x) \Rightarrow Cell(x),$
  $hasPart(x, f_3(x)), Euk.Ribosome(f_3(x)),$
  $hasPart(x, f_4(x)), Euk.Chromosome(f_4(x)),$
  $hasPart(x, f_5(x)), Nucleus(f_5(x)),$
  $inside(f_4(x), f_5(x)), f_3(x) = f_1(x), f_4(x) = f_2(x),$
  $hasRoot(f_3(x), x), hasRoot(f_4(x), x),$
  $hasRoot(f_5(x), x),$
  $pairwise\_unequal(\{x, f_3(x), f_4(x), f_5(x)\})$
$Ribosome(o_{Ribosome}), Chromosome(o_{Chromosome})$
$\ldots o_{Cell} \neq o_{Euk.Cell} \ldots$ (axiom sets 2–4 from Def. 10)

We claim that we can decide the provenance problem for the strengthened GSKB Σ' syntactically as follows; also recall that in an admissible KB, the consequents do not contain redundant concept atoms:

Definition 11. *Syntactic provenance of atoms in Σ'. In a strengthened GSKB Σ', for $C \in concepts(\Sigma)$, let $\alpha \in \tau_{C,\Sigma'}$ be an atom:*

- $\alpha = C(f(x))$ *is inherited from D if $D(f_s(x)) \in \tau_{C,\Sigma'}$ with $D \in \{C\} \cup$ all_superclasses(C, Σ') and $f'(f_s(x)) = f(x) \in \tau_{C,\Sigma'}$ with $C(f'(x)) \in \tau_{D,\Sigma'}$, and there is no more general class $SupD$ with has_superconcept$(D, SupD)$ for which this is also the case.*
- $\alpha = R(f_1, f_2)$ *is inherited from D if $D(f_s(x)) \in \tau_{C,\Sigma'}$ with $D \in \{C\} \cup$ all_superclasses(C, Σ') and $\{f'_1(f_s(x)) = f_1(x), f'_2(f_s(x)) = f_2(x)\} \subseteq \tau_{C,\Sigma'}$ with $R(f'_1, f'_2) \in \tau_{D,\Sigma'}$, and there is no more general class $SupD$ with has_superconcept$(D, SupD)$ for which this is also the case.*

If α is not inherited from some concept, it is called local *to C.*

Looking at the example GSKB Σ', we see that the atoms $hasPart(x, f_3(x))$ are inherited from $Cell$, due to $f_3(x) = f_1(x)$, and $hasPart(x, f_4(x))$, due to $f_4(x) = f_2(x)$. Consequently, $hasPart(x, f_5(x)), Nucleus(f_5(x)), inside(f_4(x), f_5(x))$ are local to $EukaryoticCell$. Hence, for the original GSKB Σ, $hasPart(x, x_3)$ and $hasPart(x, x_4)$ were inherited from $Cell$, and $hasPart(x, x_5), Nucleus(x_5), inside(x_4, x_5)$ are local to $EukaryoticCell$.

We claim that we can decide the co-reference problem for the strengthened GSKB Σ' syntactically as follows:

Definition 12. *Syntactic co-reference of terms in Σ'. Two terms with $t_1 \in terms(C, \Sigma')$, $t_2 \in terms(D, \Sigma')$ are co-referential, if $t_1 = t_2 = x$, or $t_1(x) = t_2(t) \in \tau_{C,\Sigma'}$, or $t_2(x) = t_1(t) \in \tau_{D,\Sigma'}$ (note that $t = x$, or $t = f_s(x)$).*

Looking at the example GSKB Σ', we see that $f_3(x) = f_1(x)$ are co-referential and hence the *Ribosome* in *Cell* is the same as the *Euk.Ribosome* in *EukaryoticCell*, and likewise for the *Chromosome* due to $f_4(x) = f_2(x)$.

Lemma 3. *Syntactic provenance according to Def. 11 is sound and complete for deciding semantic provenance according to Def. 8. Syntactic co-reference according to Def. 12 is sound and complete for deciding semantic co-reference according to Def. 8.*

Proof. Soundness is immediate. Completeness is a straightforward too, as Skolem functions are not shared by different consequents in Σ', and Σ was admissible. Moreover, for two different Skolem functions f_i, f_j, with $i \neq j$, $f_i(t) = f_j(t)$ will hold for a certain term t in *all* models of Σ' if and only if this was explicitly enforced by an equality atom. Note that this also proves Proposition 3.

We can generalize these results to the original GSKB Σ as follows. To check the provenance of $\tau_{C,\Sigma}$ we need to keep track during Skolemization which atom $\alpha' \in \tau_{C,\Sigma'}$ corresponds to α, and check the provenance of α' in Σ'. Likewise, to check to coreferentiality of two variables, let t_1 and t_2 be its corresponding (Skolem function) terms in the skolemized versions. Now, x_1 and x_2 are co-referential in Σ iff t_1 and t_2 are co-referential in Σ'. Looking at the example GSKB Σ, we see that x_1 from $S1$ is co-referential with x_3 in $S3$ since $f_3(x) = f_1(x)$ in Σ', and x_2 from $S1$ is co-referential with x_4 in $S3$ due to $f_4(x) = f_2(x)$ in Σ'.

However, given that a GSKB may have more than one strengthened version, "to decide" should be understood in a *credulous* way here. Only in case a provenance information or co-reference holds in *all* strengthened GSKBs, this would be a *skeptical* conclusion; it is clear that all strengthened GSKBs can in principle be constructed, due to finiteness of $\mathcal{I}_\mathcal{H}$. We can hence present the main result of this paper as follows:

Corollary 1. *Given a strengthened GSKB Σ', we can decide the provenance and co-reference problem on a syntactic basis. For an admissible (input) GSKB Σ, we can decide the* credulous *provenance and credulous co-reference problem by constructing a strengthened GSKB Σ' first, and check there. The skeptical provenance and skeptical co-reference problem can be decided by constructing all strengthened GSKBs, and checking if a positive answer holds in all of them.*

3   Graph Structured Knowledge Bases – The Graph-Based Perspective

As outlined in the Introduction, in this Section we address the provenance and co-reference problems from a graph-theoretic perspective. GSKB and CGraphs are defined as graph-based notions, and the so-called GSKB covering algorithm is presented as an algorithm which establishes graph morphisms between CGraphs. Those morphisms describe how content is inherited from other CGraphs, throughout the GSKB. The actual implementation of the algorithm is also discussed, together with implementation tricks which make it scalable.

A major drawback of the previously given logic-based framework was the disallowance of *cyclical concepts,* i.e. of concepts in which the transitive closure of the *refers to* (or *uses*) relation is not irreflexive. In the AURA GSKB many concepts are actually cyclical in that sense, for example, the concept *AnimalCell* refers to *Animal*, and vice versa. We disallowed cyclical concepts in the logic-based formalization in order to ensure the existence of *finite* Herbrand models. In the following *graph-based formalization* we *allow* cyclical concepts and describe a strategy for dealing with them. In addition, we will also show how to handle a relation hierarchy (something which could have been done in the logical formalization, too, but was omitted for the sake of brevity – the same applies to disjointness axioms and other straight-forward representation means).

Intuitively, in case of a refers-to cycle, the algorithm will make a non-deterministic choice, as illustrated by the following example. Consider it is stated in *Animal* that *Animal*s have *AnimalCell*s as parts, and conversely within *AnimalCell*, that *AnimalCell*s are *partOf Animal*s. In this case the algorithm will tell us that the fact "*Animal*s have *AnimalCell* parts" is either inherited from *Animal* to *AnimalCell*, or vice versa, but not both. Hence, a fact (atom, triple) is always owned or local to one concept.

3.1 Definitions

Definition 13. *Graph-Structured Knowledge Base. A graph-structured knowledge base (GSKB) is a tuple* $(\mathcal{C}, \mathcal{R}, \mathcal{O_C}, \mathcal{O_R}, \mathcal{I_R}, \mathcal{G})$, *where* $\mathcal{C}$ *is a set of concepts,* $\mathcal{R}$ *is a set of binary relations,* $\mathcal{O_C} \subseteq \mathcal{C} \times \mathcal{C}$ *is the concept taxonomy, and* $\mathcal{O_R} \subseteq \mathcal{R} \times \mathcal{R}$ *is the relation hierarchy. Both are strict partial orders. For a relation R, $(R, S) \in \mathcal{I_R}$ means that S is the inverse relation of R, and vice versa:* $\mathcal{I_R} = \mathcal{I_R}^{-1}$, *so $\mathcal{I_R}$ is closed under the symmetric closure. We denote the inverse of R with R^{-1} – note that for every R, either $(R, S) \in \mathcal{I_R}$, for some $R \neq S$, or $(R, R^{-1}) \in \mathcal{I_R}$. Moreover, $\mathcal{G}$ is the set of CGraphs for $\mathcal{C}$, one per concept. With $\mathcal{G}_C$ we denote the CGraph of concept C.*

Intuitively, $(C, D) \in \mathcal{O_C}^+$ means that D is a superconcept or a more general concept than C, and that it is possible that the CGraph of C contains inherited content from the CGraph of D. But note that C may also contain inherited content from non-superconcepts, which are instantiated in C. CGraphs are defined as follows:

Definition 14. *CGraph of a concept C. A CGraph of a concept C, $\mathcal{G}_C$, is a node- and edge-labeled graph* (N_C, E_C, L_C^N, L_C^E), *with nodes N_C and edges E_C. There is a special node $root_C \in N_C$, the root node. Two edges (n_1, n_2) and (n_3, n_4) are called adjoined if they share a node:* $\{n_1, n_2\} \cap \{n_3, n_4\} \neq \emptyset$. *We require that every node n is connected to the root node $root_C$:* $(root_C, n) \in E_C^{\otimes}$, *and $E_C^{\otimes}$ denotes the symmetric and transitive closure of E_C.*

The labeling functions $L_C^N : N_C \mapsto 2^{\mathcal{C}}$ and $L_C^E : E_C \mapsto 2^{\mathcal{R}}$ are total functions; labeling with $\emptyset$ is permitted.

The augmented CGraph of $\mathcal{G}_C$ is $\mathcal{A}\mathcal{G}_C$, which is a CGraph that satisfies the following:

- $E_C = E_C^{-1}$, *so E_C is closed under symmetric closure.*
- *for all $(n_1, n_2) \in E_C$, if $R \in L_C^E(n_1, n_2)$, then $R^{-1} \in L_C^E(n_2, n_1)$.*

- $L_C^N(n) = \mathcal{L}$, then $\mathcal{L}$ is closed under implied superconcepts: if $D \in \mathcal{L}$, and $(D, E) \in \mathcal{O}_C^+$, then $E \in \mathcal{L}$.
- $L_C^E(n_1, n_2) = \{S \mid R \in L_C^E(n_1, n_2), \text{ and } S = R \text{ or } (R, S) \in \mathcal{O}_{\mathcal{R}}^+\}$.

As a shorthand, we say that $F(n)$ is a concept atom in C if $n \in N_C$ and $F \in L_C^N(n)$. Analogously, we are saying that $R(n_1, n_2)$ is a relation atom in C if $(n_1, n_2) \in E_C$ and $R \in L_C^E((n_1, n_2))$. Two relation atoms $R(n_1, n_2)$ and $S(n_3, n_4)$ are adjoined if $\{n_1, n_2\} \cap \{n_3, n_4\} \neq \emptyset$.

We refer to the set of concept (relation) atoms in $\mathcal{AG}_C$ as CA_C (RA_C).

Note that a CGraph of C can be trivial, i.e. if C has no further structure, then $N_C = \{root_C\}$, $L_C^N = \{(root_C, \{C\})\}$, $E_C = \emptyset$, $L_C^E = \emptyset$. We also use the atom notation to denote graphs, e.g., $C = \{C(root_C)\}$.

3.2 Concept Graph Morphisms and Inherited Atoms

Definition 15. *CGraph Morphism from C into D. Let $\mathcal{G}_C$, $\mathcal{G}_D$ be the augmented CGraphs for C, D, with $C \neq D$. A CGraph morphism (morphism for short) from C into D, or $\mathcal{G}_C$ into $\mathcal{G}_D$, is a partial function $\mu_{C \leadsto D} : N_C \to N_D$ (denoted as μ in case C, D are clear from the context or irrelevant) such that*

1. *$\mu(root_C) = n$ is defined - n will also be called a C expansion start node.*
2. *for all $t \in N_C$ s.t. $\mu(t)$ is defined: if $L_C^N(t) = \mathcal{L}$, then $L_D^N(\mu(t)) = \mathcal{L}'$ with $\mathcal{L} \subseteq \mathcal{L}'$.*
3. *if $\mu(m_k) = n$, $m_k \neq root_C$, $k \geq 1$, then there is a sequence of adjoining relation atoms $\langle R_1(m_0, m_1), \ldots, R_k(m_{k-1}, m_k)\rangle$ in C with $m_0 = root_C$ such that there is a corresponding sequence of adjoining relations atoms in D: $\langle R_1(\mu(m_0), \mu(m_1)), \ldots, R_k(\mu(m_k), n)\rangle$ in D.*

More specifically, the morphism with $\mu_{C \leadsto D}(root_C) = n$ is denoted as $\mu_{C \leadsto D|n}$. Note that n is a node in D for which $C \in L_D^N(n)$ holds, or equivalently, $C(n) \in CA_D$, and that $\mu_{C \leadsto D|n}(root_C) = n$. If n is irrelevant, we only write $\mu_{C \leadsto D}$. C is called the domain (or source) concept, and D the range (or target) concept; we also use domain (or source) and range (or target) CGraph.

A morphism $\mu_{C \leadsto D}$ is more general than $\mu_{SubC \leadsto D}$ iff $(SubC, C) \in \mathcal{O}_C^+$.

Hence, the labels of the nodes in the source CGraph have to be subsets of the labels of the corresponding nodes in the target CGraph which are, hence, more specific (or identical). Note that the analog does *not* hold for the edges. Rather, a morphism *induces* inherited edges in the target CGraph, to be defined below.

We say that a morphism *induces atoms*:

Definition 16. *Induced / Inherited and Local Atoms. Let $\mu_{C \leadsto D|n}$ be a morphism. Then, the morphism induces the following atoms:*

- *$E(m) \in CA_D$ is induced if there is $\mu(m') = m$ with $E(m') \in CA_C$.*
- *$R(m_1, m_2) \in RA_D$ is induced if there is $\mu(m_1') = m_1, \mu(m_2') = m_2$ with $R(m_1', m_2') \in RA_C$.*

Induced atoms are also called inherited atoms. *More specifically, we say that an atom is* induced by C into D, *or* inherited from C to D. *The set of induced atoms is denoted by $\psi_{C \leadsto D|n}$. An atom in D which is not induced is called a* local atom, *more specifically:* local to D.

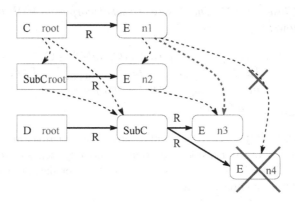

Fig. 2. Illustration of closure and inconsistency

3.3   Concept Patchworks and Coverings

Obviously, a complex CGraph may have more than one concept atom, and content may be inherited from any of those. Thus, *multiple morphisms* are required to describe all inherited content. For example, if D contains $C(n_1)$ and $C(n_2)$ with $n_1 \neq n_2$, we already require two morphisms: $\mu_{C \rightsquigarrow D|n_1}, \mu_{C \rightsquigarrow D|n_2}$ (μ is a function). A morphisms is hence already associated with the *expansion start node*, here n_1 and n_2. A *CGraph patchwork* describes how *different parts of the concept are "patched" together*:

Definition 17. *CGraph patchwork into D, induced atoms, unified nodes. A CGraph patchwork (patchwork for short) into D for a subset of its concept atoms $T \subseteq CA_D \setminus \{D(root_D)\}$ is a set of morphisms into D, $\mathcal{PW}_D \triangleq \{\mu_{E \rightsquigarrow D|n} \mid E(n) \in T\}$.*

The set of induced or inherited atoms is $\psi(\mathcal{PW}_D) = \bigcup_{\mu \in \mathcal{PW}} \psi_\mu$.

Let $\mathcal{I}_\alpha = \{\mu \mid \alpha \in \psi_\mu, \mu \in \mathcal{PW}_D\}$ be the set of morphism that induces the atom α. We say that α is inherited by $\mu_{C \rightsquigarrow D} \in \mathcal{I}_\alpha$ if $\mu_{C \rightsquigarrow D}$ is a (or the) most general morphism in $\mathcal{I}_\alpha$. Given $\mathcal{PW}_D$, we hence say that α is inherited from C.

If $\{\mu_{E_1 \rightsquigarrow D}, \mu_{E_2 \rightsquigarrow D}\} \subseteq \mathcal{PW}_D$ with $\mu_{E_1 \rightsquigarrow D}(n_1) = \mu_{E_2 \rightsquigarrow D}(n_2) = n$, then we say that n_1 and n_2 have been unified (merged, equated) into n, and n_1 and n_2 are said to be co-referential. Note that possibly $E_1 = E_2$.

The *closure of a set of morphisms* is defined as some kind of transitive closure over the different morphisms. Consider the concepts C, $SubC$, and D, with $(SubC, C) \in \mathcal{O}_C^+$, and the morphisms $\mu_{C \rightsquigarrow SubC}, \mu_{SubC \rightsquigarrow D}$ with $\mu_{C \rightsquigarrow SubC}(root_C) = root_{SubC}$, hence $\mu_{C \rightsquigarrow SubC|root_{SubC}}$ (note that $C(root_{SubC})$ is a concept atom in $SubC$, because of $(SubC, C) \in \mathcal{O}_C^+$!), and $\mu_{SubC \rightsquigarrow D}(root_{SubC}) = n_{SubC}$, hence $\mu_{SubC \rightsquigarrow D|n_{SubC}}$. Moreover, assume that $\mu_{C \rightsquigarrow SubC|root_{SubC}}(n_1) = n_2$. Then, if also $\mu_{SubC \rightsquigarrow D|n_{SubC}}(n_2) = n_3$ holds, we have to ensure that $\mu_{C \rightsquigarrow D|n_{SubC}}(n_1) = n_3$ holds (note that $C(n_{SubC})$ is a concept atom in D). Suppose to the contrary that also $\mu_{C \rightsquigarrow D|n_{SubC}}(n_1) = n_4$, with $n_3 \neq n_4$. This describes a situation where n_1 was inherited directly and indirectly from C into D, and gets duplicated as n_3, n_4. The situation is depicted in Fig. 2. We hence require that the composition of the morphisms does not violate functionality of the morphisms.

Definition 18. *Closure of $\mathcal{PW}$. The closure of a set of morphisms $\mathcal{PW}$, denoted $\mathcal{PW}^+$, is defined as follows:*

- $con(\mathcal{PW}) \triangleq \{\mu_{C \rightsquigarrow D|n_{SubC}} \cup \{(n_1, n_3)\} \mid (C, SubC) \in \mathcal{O}_C{}^+,$
  $\{\mu_{C \rightsquigarrow SubC|root_{SubC}}, \mu_{SubC \rightsquigarrow D|n_{SubC}}\} \subseteq \mathcal{PW},$
  $\mu_{C \rightsquigarrow SubC|root_{SubC}}(n_1) = n_2, \mu_{SubC \rightsquigarrow D|n_{SubC}}(n_2) = n_3\}$
- $\mathcal{PW}^+ \triangleq \bigcup_{i \in 1\ldots\infty} con^i(\mathcal{PW})$

(Note that $(D, SubC) \in \mathcal{O}_C{}^+$ is possible, also.) The relations in $\mathcal{PW}^+$ are now no longer necessarily functions, since we extended them by means of $\mu_{C \rightsquigarrow D|n_{SubC}} \cup \{(n_1, n_3)\}$, i.e., we added $\mu_{C \rightsquigarrow D|n_{SubC}}(n_1) = n_3$. However, the notion of *consistency of a set of morphisms* requires exactly this:

Definition 19. *Consistency of $\mathcal{PW}$. A set of morphisms $\mathcal{PW}$ is called* consistent *iff, for all C, D, n, every morphism in $\mathcal{PW}^+$ satisfies the following:*

1. *functionality: if $\mu_{C \rightsquigarrow D|n}(n_1) = n_2 \in \mathcal{PW}^+$ and $\mu_{C \rightsquigarrow D|n}(n_1) = n_3 \in \mathcal{PW}^+$, then $n_2 = n_3$.*
2. *no self mapping: for all $C, n, n_1 \neq n_2, \mu_{C \rightsquigarrow C|n}(n_1) = n_2 \notin \mathcal{PW}^+$.*
3. *no self inheritance: there is no sequence of morphisms $\mu_1, \ldots, \mu_m \in \mathcal{PW}^+, m > 1$ such that $(\mu_1 \circ \cdots \circ \mu_m)(n) = n$ (where $\circ$ denotes composition).*

The functionality criterion has already been explained. The second criterion prevents that a CGraph C inherits content to itself, which is a reasonable assumption.[3] The third criterion simply prevents that a node is inherited from itself, directly or indirectly.

Our goal is to identify the inherited content in a GSKB by means of patchworks, for all its concepts. Often, there is more than one possible patchwork for a given concept C, and different patchworks may result in different sets of inherited / induced atoms ψ. We are interested in those patchworks that *globally maximize (for the whole GSKB)* the set of inherited atoms ψ, and among those the *smallest* such patchworks, i.e., those that require as few morphisms as possible. In case an atom can be inherited via two different morphisms such that one is more general than the other, preference should be given to the more general morphism. This captures the intuition that an atom should be inherited from the most general concept that has it. These ideas are formalized in the notion of a *GSKB covering* as follows:

Definition 20. *GSKB Covering. A covering of a GSKB K is a union of CGraph patchworks for a (not necessarily true) subset of its concepts $C' \subseteq C$: $\mathcal{PW}_K \triangleq \bigcup_{C \in C'} \mathcal{PW}_C$. We extend the definitions of induced / implied atoms to whole knowledge bases as follows: $\psi(\mathcal{PW}_K) \triangleq \bigcup_{\mathcal{PW}_C \in \mathcal{PW}_K} \{(C, \alpha) \mid \alpha \in \psi_C(\mathcal{PW}_C)\}$. We then require that $\mathcal{PW}_K$ a minimal set (w.r.t. set inclusion) that satisfies the following three principles:*

1. *consistency: $\mathcal{PW}_K$ is consistent.*
2. *ψ-maximality: there is no GSKB covering $\mathcal{PW}'_K$ of the same or smaller cardinality as $\mathcal{PW}_K$ with $\psi(\mathcal{PW}_K) \subset \psi(\mathcal{PW}'_K)$.*

---

[3] Even if C was cyclical in the sense that $C(n)$ is a concept atom in C, for $n \neq root_C$.

3. μ-generality: if $\alpha \in \psi_{C \leadsto D} \in \mathcal{PW}_K$ for some C, D, and there exists a more general morphisms $\psi_{C' \leadsto D}$ with $\alpha \in \psi_{C' \leadsto D}$ with $(C, C') \in \mathcal{O}_C^+$, such that its addition to $\mathcal{PW}_K$ would retain consistency of $\mathcal{PW}_K$, then $\psi_{C' \leadsto D} \in \mathcal{PW}_K$.

In general, there can be more than one covering for a given GSKB. An example is given by the cyclical GSKB with concepts C, D, such that $C = \{C(root_C), R(root_C, n_1), D(n_1)\}$, and $D = \{D(root_D), R^{-1}(root_D, n_2), C(n_2)\}$. Here, we can either have a morphisms $\mu^1_{C \leadsto D|n_2}(root_C) = n_2, \mu^1_{C \leadsto D|n_2}(n_1) = root_D$, or a morphisms $\mu^2_{D \leadsto C|n_1}(root_D) = n_2, \mu^2_{D \leadsto C|n_1}(n_2) = root_C$, but not both. Hence, a covering forces that one of the two relation atoms in C, D is local, and the other one is inherited. If both were inherited, then point point 3 in Def. 19 would be violated. If both were local, then the principle of ψ-maximality would be violated. It is left unspecified which of those two coverings is constructed by the algorithm given below. Suppose we chose $\mu^1_{C \leadsto D}$ for $\mathcal{PW}_K$. Note that we cannot even add $\mu^3_{D \leadsto C} = \{(root_D, n_1)$ to $\mathcal{PW}_K$, as this would result in $(\mu^1_{C \leadsto D} \circ \mu^3_{D \leadsto C})(n_1) = n_1$ and $(\mu^3_{D \leadsto C} \circ \mu^1_{C \leadsto D})(root_D) = root_D$, hence $D(root_D)$ in D and $D(n_1)$ in C would become self-inherited, hence violating consistency again.

The principle of μ-generality is explained as follows. Consider the CGraphs C, C', C'', such that $\{(C', C), (C'', C')\} \subseteq \mathcal{O}_C$, with $C = \{C(root_C), R(root_C, n_1), D(n_1)\}$, $C' = \{C'(root_{C'}), R(root_{C'}, n_2), D(n_2), R(n_2, n_3), E(n_3)\}$ and $C'' = \{C''(root_{C''}), R(root_{C''}, n_4), D(n_4), R(n_4, n_5), E(n_5), R(n_5, n_6), F(n_6)\}$. Looking at C'', the atoms $R(root_{C''}, n_4), D(n_4), R(n_4, n_5), E(n_5)$ can all be induced by C' into C'' via the obvious morphism.[4] However, $R(root_{C''}, n_4), D(n_4)$ can already be induced by C into C', and this is the more general morphism.[5] We hence require that this morphism is also present: $\mu_{C \leadsto C''} \in \mathcal{PW}_K$, since those atoms are induced by both morphisms. Also, we need $\mu_{C' \leadsto C''} \in \mathcal{PW}_K$ as otherwise, we would not be able to inherit $R(n_4, n_5), E(n_5)$ from C' to C'', hence violating ψ-maximality. The desired covering will hence consider the following atoms as local: all atoms in C, the atoms $R(n_2, n_3), E(n_3)$ in C', and the atoms $R(n_5, n_6), F(n_6)$.

3.4   Computation of a Covering

Our goal is to compute a GSKB covering, and it is clear that this is a hard problem:

Proposition 4. *Computing a Covering is NP-Hard and in PSPACE. It is clear that we can compute a covering of K in a simple non-deterministic "guess and validate" style – we guess the morphisms, and check whether they satisfy the conditions Def. 19 and Def. 20. However, storing the morphisms may require polynomial space. It is also clear that the problem is NP-hard, by reduction from the clique problem: let G be the input graph to the clique problem, and the CGraph of $SupG$ be a k-clique, and $(G, SupG) \in \mathcal{O}_C^+$. Assume $L_{SupG}(n) = \emptyset$ for all $n \in N_{SupG}$, $n \neq root_{SupG}$, and $L_{SupG}(n) = \emptyset$ for all $n \in N_G$ for $n \neq root_G$, and $L_G(root_G) = \{G, SupG\}, L_{SupG}(root_{SupG}) = \{SupG\}$, with $C = \{SupG, G\}$. Then, G contains a clique of size k iff, for its covering $\mathcal{PW}_K$: $\psi_G(\mathcal{PW}_K) = CA_G \cup RA_G \setminus \{SupG(root_G)\}$ (i.e., all relation and concept*

---

[4] Be means of $\mu_{C' \leadsto C''}(root_{C'}) = root_{C''}, \mu_{C' \leadsto C''}(n_2) = n_4, \mu_{C' \leadsto C''}(n_3) = n_5$.

[5] By means of $\mu_{C \leadsto C''}(root_C) = root_{C''}, \mu_{C \leadsto C''}(n_1) = n_4$.

atoms with the exception of the root concept atom are inherited), and all morphisms are injective.

The deterministic version of the sketched nondeterministic algorithm requires search. The basic ideas for a deterministic version of the algorithm are outlined as follows.

Since the GSKB can contain cyclical references, we must prevent the construction of morphisms which induce cycles in order to satisfy the requirements of Def. 19, namely points 2 and 3. In principle, we could compute the patchworks for the concepts in any order of sequence. When the patchwork for concept D is constructed, we are establishing maximal morphisms to all concepts mentioned in concept atoms in D, and we make sure that we do not violate consistency, ensuring ψ-maximality. Moreover, for every concept atom $C(n)$ in D, if $(C, SupC) \in \mathcal{O}_C{}^+$, then also $SupC(n)$ in D, and hence, we will not only establish $\mu_{C \rightsquigarrow D}$, but also $\mu_{SupC \rightsquigarrow D}$, hence satisfying μ-generality.

Unfortunately, checking the conditions in Def. 19 to ensure consistency in Def. 20 can be costly, and we hence propose a slightly optimized version which works in *two phases* by exploiting ideas akin to topological sorting w.r.t. a *refers_to*$^+$ relation:

Definition 21. *Cycles, Refers-To Relation, Refers-To Violation. A concept C refers to concept D, iff $D(n) \in CA_C$: refers_to(C, D). A concept C is cyclical iff refers_to$^+(C, C)$. A GSKB K is cyclical if it contains a cyclical concept. Given two concepts C_i, C_j with $1 \le i < j \le m$ in a sequence of concepts $\langle C_1, \ldots, C_m \rangle$, we speak of a refers-to violation iff refers_to(C_i, C_j) holds. The number of refers-to violations for the sequence is ref_vio$(\langle C_1, \ldots, C_m \rangle) \triangleq \Sigma_{1 \le i \le m} |dom(refers\_to^+(C_i)) \cap \{C_{i+1}, \ldots, C_m\}|$. An optimal concept processing sequence is a sequence seq $= \langle C_1, \ldots, C_m \rangle$, $C = \{C_1, \ldots, C_m\}$, which minimizes the number of refers_to violations: $\xi = argmin(ref\_vio(seq), seq \in \pi(C))$; π denotes all possible sequences / permutations of C.*

In case $\xi = 0$, we can simply topologically sort the GSKB w.r.t. the *refers_to*$^+$ relation and process in that order of sequence. For such GSKB we only need to check point 1 in Def. 19. And, even with $\xi > 0$, we can skip the checks for point 2 and 3 in Def. 19 if we organize the computation of the covering in *two phases:* In *phase one*, after having computed an optimal sequence, to construct a patchwork for concept C_k we consider only morphisms from concepts C_i into C_k with $i < k$ w.r.t. that ordering, hence, C_i was processed before. We make sure that patchworks do not violate point 1 in Def. 19. The result of phase 1 is a covering in case $\xi = 0$. In case $\xi > 0$, ψ-maximality is not yet satisfied. There is (at least) a concept C_k with $refers\_to^+(C_k, C_l)$ and $l > k$. Hence, C_k may contain inherited content from C_l, but $\mu_{C_l \rightsquigarrow C_k}$ was not established. Those concepts C_k are identified and collected in phase 1, and simply re-processed again in *phase 2*, by re-computing their patchworks, leading to $\mu_{C_l \rightsquigarrow C_k}$. During morphisms construction, we now have to checks all points in Def. 19, which is more costly. However, the number of concepts to re-consider in phase 2 is usually much smaller than $|\mathcal{C}|$. This is Algorithm 1, which uses Algorithm 2 for patchwork computation.

The function $most\_specific\_atom(CA)$ is non-deterministic and chooses a concept atom $D(n) \in CA$ for which there is no concept atom $SubD(n) \in CA$ with $(SubD, D) \in \mathcal{O}_C{}^+$. Within $compute\_patchwork(C, \mathcal{PW}, \mathcal{P}_K, phase)$, we make sure

Input : A GSKB $K = (\mathcal{C}, \mathcal{R}, \mathcal{O_C}, \mathcal{O_R}, \mathcal{I_R}, \mathcal{G})$
Output: $\mathcal{PW}_K$: a covering of K

$\mathcal{PW}_K \leftarrow \emptyset$;
$\langle C_1, \ldots, C_n \rangle \leftarrow argmin(ref\_vio(seq), seq \in \pi(\mathcal{C}))$;
$\mathcal{S} \leftarrow \emptyset$ % concepts to reconsider in Phase 2 ;
$\mathcal{P} \leftarrow \emptyset$ % already processed concepts ;
% Phase 1 ;
for $i \leftarrow 1$ **to** n **do**
    if $dom(refers\_to^+(C_i)) \cap \{C_{i+1}, \ldots, C_n\} \neq \emptyset$ **then**
        $\mathcal{S} \leftarrow \mathcal{S} \cup \{C_i\}$;
    end
    $\mathcal{PW}_K \leftarrow \mathcal{PW}_K \cup compute\_patchwork(C_i, \mathcal{PW}_K, \mathcal{P}, 1)$;
    $\mathcal{P} \leftarrow \mathcal{P} \cup \{C_i\}$;
end
% Phase 2 ;
for $C \in \mathcal{S}$ **do**
    $\mathcal{PW}_K \leftarrow \mathcal{PW}_K \setminus \{\mu_{D \rightsquigarrow C} \mid \mu_{D \rightsquigarrow C} \in \mathcal{PW}_K\}$;
    $\mathcal{PW}_K \leftarrow \mathcal{PW}_K \cup compute\_patchwork(C, \mathcal{PW}_K, \mathcal{P}, 2)$;
end
return $\mathcal{PW}_K$

Algorithm 1. $cover(k)$ – The GSKB Covering Algorithm

Input : A concept, a set of morphisms, a set of processed concepts, a *phase* $\in \{1, 2\}$:
    $(C, \mathcal{PW}, \mathcal{P}_K, phase)$
Output: A C-patchwork $\mathcal{PW}_C$

$\mathcal{S} \leftarrow \{D(n) \mid D \in \mathcal{P}, D(n) \in CA_C\}$;
$\mathcal{PW}_C \leftarrow \emptyset$;
while $\mathcal{S}$ **do**
    $D(n) \leftarrow most\_specific\_atom(\mathcal{S})$;
    $\mathcal{S} \leftarrow \mathcal{S} \setminus \{D(n)\}$;
    $(\mu_{D \rightsquigarrow C|n}, \mathcal{P}^+) \leftarrow max\_consist.\_morphism(C, D(n), \mathcal{PW}_K \cup \mathcal{PW}_C, phase)$;
    $\mathcal{PW}_C \leftarrow \mathcal{PW}_C \cup \{\mu_{D \rightsquigarrow C|n}\} \cup \mathcal{P}^+$;
end
return $\mathcal{PW}_C$

Algorithm 2. $compute\_patchwork(C, \mathcal{PW}, \mathcal{P}_K, phase)$ – Helper Function

to process the concept atoms in CA_C in *most specific first* order. This has the advantage that the closures can be more easily computed: at the time when $\mu_{SupC \rightsquigarrow SubC}$ is computed, the morphisms $\mu_{SupC \rightsquigarrow C}$ and $\mu_{C \rightsquigarrow SubC}$ are already available.

The function $max\_consistent\_morphism(C, D(n), \mathcal{PW}_K \cup \mathcal{PW}_C, phase)$ finds the maximal consistent morphism from D into C. In case $phase = 1$, it only checks point 1 in Def. 19 in the consistency check, otherwise all points in Def. 19. It tries to compute a maximal consistent morphism which induces as many atoms as possible into C. In case no consistent mapping can be found, the empty set is returned. In order to check consistency, $max\_consistent\_morphism$ has to compute the closure of the given morphism $\mathcal{PW}_K \cup \mathcal{PW}_C$. If a maximally consistent morphism $\mu_{D \rightsquigarrow C|n}$ can be found, it is returned together with the additional morphisms resulting from the closure computation: $closure(\mathcal{PW}_K \cup \mathcal{PW}_C \cup \{\mu_{D \rightsquigarrow C|n}\}) \setminus \mathcal{PW}_K$. This function is not further detailed here, but its implementation described below.

3.5   The Implementation

The so-far presented algorithm is an idealized version of the actual implemented algorithm. During our implementation efforts, we learned that the idealized algorithm is not able to compute a covering for the AURA KB in a reasonable amount of time. Scalability is of concern here. In the AURA KB, we have 695 concepts with more than 20 nodes, on average 50.1 nodes and 104.5 edges. The average number of concept atoms in those concepts is 104.7. That means we have to consider potentially 104.7×695 graph morphisms. If runtime is exponential in the number of nodes, this results in 8.78×10^{19} node mappings to consider if a naive approach is followed, all of which have to be checked for consistency, etc. A lot of effort has been put into implementing a more scalable version of $compute\_patchwork$.

In the following we describe the main *implementation techniques that enabled scalability*; we think that it might be insightful to document and preserve those techniques for researchers working on similar problems.

First, there is no computation of an optimal sequence – rather, the optimal sequence is constructed iteratively / incrementally during phase 1.

More importantly, instead of finding a maximal morphism for each $D(n) \in CA_C$ at a time, the algorithm constructs and patches morphisms together in a piecewise fashion. It is agenda-driven – the agenda consists of the currently non-induced edges of C. Such an edge is called *open*. In a loop, an open edge $R(n_1, n_2)$ is selected from the agenda, and it is then searched for a morphism which induces it. Whenever a node mapping is established, we are checking the consistency conditions as required, and prevent cycles, etc. When the agenda is empty, i.e., there are either no more open edges or no more additional morphisms can be found, then a solution was found. The quality of the computed solution / patchwork is determined by the number of its induced atoms – *the score of the solution*.

The implemented algorithm produces perfect coverings if given indefinite time. Otherwise, the quality of the approximation increase the more time it is given (there is a timeout switch, see below). After a solution has been found, the algorithm can continue the search for additional solutions with higher scores. Only the best solutions are kept. During this search for better solutions, the score of the best so-far found solution is used to reduce the search space – sometimes work on a partial solution can be stopped if it can be determined that the full solution based on it cannot outperform the best so-far found solutions.

We can parameterize the algorithm such that the search in $compute\_patchwork$ terminates as soon as a total number of m_1 solutions has been found, or if we were able to improve the best-so-far solution m_2 times, or if a timeout after t seconds has been reached. In case the timeout occurred before a (suboptimal) patchwork was found, we can either continue the search until at least one patchwork is available, or accept the incomplete partial patchwork (hence, all remaining non-induced atoms as local).

We have three **strategies a), b), and c)** for finding a morphism for $R(n_1, n_2)$:

Strategy a) We can often find a morphism starting from $n = n_1$ or $n = n_2$, for some concept atom $D(n)$. In case there is more than one such $D(n)$, we are trying the *most specific concept atoms first*. We inspect the CGraph of D and try to inherit a path of edges, starting from the $root_D$, hence producing a morphism $\mu_{D \leadsto C|n}$. Each inherited path and hence morphism μ is associated with a *score* which is the product $|dom(\mu)| \times (penalty + 1)$. The *penalty* is the number of node and edge label specializations required over D. Ideally, we are looking for a perfect match with penalty 0 and maximal size. However, since this "path harvesting" already requires graph search in D in order to find the inherited path, the max. path length is constrained to a small number, say 3. The inherited edges are now no longer open.

Note that an edge can be induced by more than one morphism, and every $D(n)$ will eventually be considered if it has to. E.g., consider $C = \{C(root_C), R(root_C, n_1), S(n_1, n_2)\}$, $C' = \{C'(root_{C'}), R(root_{C'}, n_3), T(n_3, n_4)\}$, and $C'' = \{C''(root_{C''}), R(root_{C''}, n_5), S(n_5, n_6), T(n_5, n_7)\}$, with $\{(C', C), (C'', C')\} \subseteq \mathcal{O}_C$. Here, $\{C''(root_{C''}), C'(root_{C''}), C(root_{C''})\} \subseteq CA_{C''}$. To induce $S(n_5, n_6)$ we require $\mu_{C \leadsto C''|root_{C''}}$, and to induce $T(n_5, n_7)$ we require $\mu_{C \leadsto C'|root_{C''}}$. Note that $R(root_{C''}, n_5)$ is induced by both morphisms.

Strategy b) Here it is checked whether $R(n_1, n_2)$ can be induced by *extending* an existing morphism; this is useful because Strategy a) has a length cut-off, as described. If we find that for $R(n_1, n_2)$ there is an edge $S(n_1, n_3)$ which is already induced by a morphism $\mu_{D \leadsto C|n}$, we then restart the search for n in C by looking at D again. In case there is more than one such morphism, we select the morphism with the highest score. We skip forward to the node n_1' with $\mu_{D \leadsto C|n}(n_1') = n_1$ and try to extend the morphism. This way, we can inherit arbitrary graph structures from D, or simply extend the path to become longer, by patching together and extending partial morphisms for *one inherited path at a time*. Restarting has the advantage that Strategy a) does not invest a lot of time constructing longer and longer paths in a depth-first fashion with increasing penalties, hence the max. length is constrained to a small value. By restarting with the partial morphism which have the highest score, we achieve a kind of best-first search effect (similar to beam search), because the control is given back quickly to the agenda, so search can re-focus quickly on the most promising partial morphisms to extend.

For example, consider $C = \{C(root_C), R(root_C, n_1), S(root_C, n_2), T(n_1, n_2)\}$ and isomorphic $C' = \{C'(root_{C'}), R(root_{C'}, n_3), S(root_{C'}, n_4), T(n_3, n_4)\}$ with $(C', C) \in \mathcal{O}_C$. We can first find $\mu_{C \leadsto C'}$ which induces $\{C'(root_{C'}), R(root_{C'}, n_3), S(root_{C'}, n_4)\}$, and then restart and extend such that $\{T(n_1, n_2)\}$ is induced by $\mu_{C \leadsto C'}$.

Strategy c) If Strategies a) and b) fail for an edge edge $R(n_1, n_2)$, then, for it to be induced, it must be induced by a morphism which already maps to some node n with $E(n) \in CA_C$ such that n is connected to n_1, n_2, but n has not been considered yet as an expansion start node. To find such a morphism, we consider all morphisms that induce $E(n)$ such that n is connected to n_1, n_2, and start the search in E at $root_E$ such that $\mu_{E \leadsto C|n}(root_E) = n$ holds.

For example, consider $C = \{C(root_C), R(root_C, n_1), S(root_C, n_2), T(n_1, n_2)\}$, $C' = \{C'(root_{C'}), R(root_{C'}, n_3), S(root_{C'}, n_4), T(n_3, n_4), U(n_4, n_5), D(n_3)\}$, and $D = \{D(root_D), T(root_D, n_6), U(n_6, n_7)\}$ with $(C', C) \in \mathcal{O}_C$. The edge $U(n_4, n_5)$ in C' can obviously not be induced by the morphism $\mu_{C \rightsquigarrow C'}|root_{C'}$. Rather, we need to establish $\mu_{D \rightsquigarrow C'}|n_3 = \{(root_D, n_3), (n_6, n_4), (n_7, n_5)\}$. Note that the edge $T(n_3, n_4)$ is induced by $\mu_{C \rightsquigarrow C'}|root_{C'}$ as well as by $\mu_{D \rightsquigarrow C'}|n_3$. It is interesting to note that this edge is hence *inherited from both C and D*, as both are incomparable w.r.t. $\mathcal{O}_C{}^+$.

4 Evaluation

We have applied the GSKB covering algorithm to the AURA GSKB, and an approximate covering with sufficient quality was computed in 18 hours. We used a timeout of 3 minutes, and after the timeout the algorithm was allowed to produce at least one solution (hence $t = 300$, and $m_1 = 50$, and $m_2 = 3$). Only 1431 concepts (22 % of the concepts) had to be re-processed in phase 2, and 20 concepts timed out – three concept required 34, 19, and 15 minutes, resp. It should be noted that the algorithm can compute approximate coverings of lower quality in much less than 18 hours, e.g. in about 2 to 3 hours.

The following table shows the stats of the covering produced by the 18 hours run, for concepts which have at least $\{0, 20, 50, 100, 200\}$ nodes ($|N_C| \geq$), and the number and size of their morphisms, as well as the percentages of inherited concept and relation atoms. It can be seen that the algorithm has performed a non-trivial task. For $|N_C| \geq 50$, 5690 morphism have been computed, with an average size of $|dom(\mu)| = 7.36$ nodes. If we look at the biggest concept with 461 nodes, we find that the algorithm has created a patchwork with 126 morphisms, with an avg. size of 9.49 nodes. Altogether, the algorithm has established $14, 146$ morphisms of average size 5.47, and identified 57 % of the concept atoms and 69 % of the relation atoms as inherited. We are also able to tell from *where* an atom is inherited from, see Def. 17. This information can be called the *provenance* of atoms and is of great importance to the modelers of the KB [4].

$|N_C|$	≥ 0	≥ 20	≥ 50	≥ 100	≥ 200
concepts (#)	6430	695	224	57	3
avg. $|N_C|$ (# nodes)	8.2	50.1	88.46	139.52	342
avg. $|E_C|$ (# edges)	14.5	104.5	198.22	328.29	866.33
avg. $|CA_C|$ (# atoms)	26.9	104.68	158.125	238.23	546
% inherited $|RA_C|$	69	75.6	77.4	77	71.6
% inherited $|CA_C|$	57	71.1	74	74.6	74.9
$|\mathcal{PW}_K|$ (# morphisms)	14146	9900	5690	2264	287
avg. $|dom(\mu)|$ (# nodes)	5.47	6.71	7.36	7.73	8.25
avg. $|\mathcal{PW}_C|$ (# morphisms)	2.2	14.24	25.4	39.71	95.67

5 Related Work

As stated in the introduction, the co-reference problem has been studied to some extent in the natural language processing literature under the the term anaphora resolution; for example, [5,6] use default rules to hypothesize equality assertions between referents in order to guess and establish co-references.

The reasoning system *Knowledge Machine (KM)*, [9], uses a so-called unification mechanism, Umap, to strengthen the GSKB by establishing co-references. Unfortunately, the lack of a formal semantics makes it very difficult to understand. A major problem in KM is that unifications are not reversible, since they are not represented explicitly as (equality or Umap) atoms in the KM GSKB. Instead, unifications are performed by destructive substitutions of constants in the GSKB. Retraction and comprehension of Umap unifications can be very difficult and time consuming, as unification is heuristic in nature and frequently goes wrong.

KM provides a function `get-supports` for computing provenance of every triple in a GSKB. The function returns the concept from where the triple gets inherited. This function relies on KM's so-called *explanation database*, which became corrupted during the AURA project, due to software errors and a changing semantics, hence forcing us to recompute the provenance information. This was the main motivation for the work described in this paper. It turned out that the recomputed provenance information was quite accurate, as confirmed by the experts in our knowledge factory [4].

The work of [10] uses answer set programming (ASP) to formalize KM's Umap operator. The GSKB is specified as an ASP program, together with an axiomatic system of ASP inference rules. These rules capture the semantics of object-oriented knowledge bases, including inheritance, and formalize the UMap operator. The semantics is given by the semantics of the ASP rules, whereas our approach starts with a notion of desirable models and is hence more model-theoretic in nature. Moreover, constants are distinct by definition in ASP programs, so equality (UMap) needs to be modeled on the meta-level. Moreover, the approach has not yet been applied successfully to the full-scale AURA GSKB, so scalability of the approach is still open.

Considering our graph-based approach, we note that graph morphism-based approaches were employed since the early days of KL-ONE [11], to decide subsumption between concepts in description logics. Those approaches are called *structural subsumption algorithms* nowadays [1]. Note that we do not decide subsumption between concepts, as the taxonomy is considered given and fixed here. Instead, we determine for a (possibly singleton) set of atoms from a concept from where they got inherited, or if they are local to that concept, and hence non-redundant. Determining from where an atom is inherited was called provenance computation. Nevertheless, we suspect that our algorithm could be turned into a structural subsumption checker for certain description logics, similar as in [12] for $\mathcal{EL}$. So-called *simulations* are employed in [13] to decide concept subsumption in $\mathcal{EL}^{++}$, which are similar to morphisms. But note that, unlike $\mathcal{EL}$, we are supporting graph-based descriptions, and this is a key requirement in AURA.

Morphisms are also considered in [14] for simple conceptual graphs for deciding "projection" (a form of subsumption). However, no implementation is described there, and we are using different morphisms, as we do not require that *all* relation atoms in C have to be projected into D by a morphism $\mu_{C \leadsto D}$.

The graph-based algorithm was very successful in the AURA project. We were not able to achieve similar results with any other form of formal reasoning, i.e., description logic reasoners. As argued, in order to compute provenance of atoms / triples, one needs a form of hypothetical, unsound reasoning which requires guessing of co-references /

equalities. Considering the size of the AURA GSKB and the complexity of the problem, we consider our algorithm a success story.

It is well-known that the modeling of graph structures is challenging in description logic (DL), as derivations from the tree-model property usually result in decidability problems [15] which can often only be regained by imposing severe artificial modeling restrictions. Although some progress has been made on modeling graph structures with DLs [16], those extensions are still too restricted to be useful here. Our experience is that graph structures are central to biology, and approximating them by trees results in coarse models. Our framework allows us to express the graph structures truthfully, but comes with other restrictions, too. To the best of our knowledge, there is no body of work in the DL community that provides answers to the problems addressed in this paper, and we are not aware of any abduction or hypothesization algorithm which has ever successfully been applied to a GSKB of this size.

We also claim that the algorithm and its implementation techniques can be applied in related areas, for example, for ontology alignment tasks in graph-structured lightweight ontologies, for applications in computational biochemistry (identification of chemical substructures), in text understanding, and in the semantic web (e.g., identification of substructures in RDFS triple stores). This is not surprising, since the MSC problem is a very general one, which shows up in many disguises in many application contexts.

6   Conclusions and Outlook

We showed how to identify inherited content in graph-structured knowledge bases, and did this from two perspectives. From a logical perspective, we argued that accurate provenance of atoms requires identification of / the proper solution of the co-reference problem. We demonstrated that inheritance structures can be captured by means of Skolem function and equality atoms. We described a so-called KB-strengthening algorithm which guesses / hypothesizes co-references between Skolem function terms in order to maximize the inherited content in concept graphs.

For the actual implementation, we employed graph-based notions and demonstrated how inherited content can be described by virtue of graph morphisms. The implemented algorithm was successfully applied to the large-scale AURA KB.

The AURA system and the actual implementation of the algorithm covers additional expressive means that we have not formally reconstructed yet, i.e., transitive and functional relations, number restrictions, and disjointness axioms. The logical formalization of these expressive means is future work, but we are optimistic that it can be done. Especially in the case of cyclical GSKBs, it is not clear yet how to apply a similar line of argumentation, as the Herbrand models are no longer necessarily finite. However, it is in principle clear how to handle disjointness, functionality etc. from a graph-based point of view.

From the morphisms computed by the graph-based algorithm we can compute a strengthened GSKB. The established equality axioms between the Skolem function terms describe the correct inheritance structures, and the quality of the mappings got testified by the subject matter experts in our knowledge factory.

The strengthened GSKB in first-order logic is also the basis for a couple of AURA knowledge base exports in SILK [17], answer-set programming syntax [18], and TPTP

FOF syntax [19]. We also have a set of OWL2 [20,21] exports [22], but the OWL2 KBs are underspecified in the sense that we cannot use Skolem functions here and hence, no equality atoms can be used to establish co-references, so those KBs are approximations. This suite of exported KBs is called the *Bio_KB_101* suite, and is made available to the public under a *Creative Commons License,* and can be found here for download [23], together with a detailed description of the various variants of the exports.

In future work, we need to establish a closer correspondence between the logical and the graph-based formalizations. It should be relatively straight forward to show that the graph-based algorithm is sound and complete w.r.t. the given logical semantics.

Acknowledgments. This work was funded by Vulcan Inc. under Project Halo.

References

1. Baader, F., Calvanese, D., McGuinness, D.L., Nardi, D., Patel-Schneider, P.F. (eds.): The Description Logic Handbook: Theory, Implementation, and Applications. Cambridge University Press (2003)
2. Gunning, D., Chaudhri, V.K., et al.: Project Halo Update - Progress Toward Digital Aristotle. AI Magazine (2010)
3. Reece, J.B., Urry, L.A., Cain, M.L., Wasserman, S.A., Minorsky, P.V., Jackson, R.B.: Campbell Biology, 9th edn. Pearson Education, Harlow (2011)
4. Chaudhri, V., Dinesh, N., et al.: Preliminary Steps Towards a Knowledge Factory Process. In: The Sixth International Conference on Knowledge Capture (2011)
5. Carpenter, B.: Skeptical and Credulous Default Unification with Applications to Templates and Inheritance. In: Inheritance, Defaults and the Lexicon, pp. 13–37. Cambridge University Press (1994)
6. Cohen, A.: Anaphora Resolution as Equality by Default. In: Branco, A. (ed.) DAARC 2007. LNCS (LNAI), vol. 4410, pp. 44–58. Springer, Heidelberg (2007)
7. Overholtzer, A., Spaulding, A., Chaudhri, V.K., Gunning, D.: Inquire: An Intelligent Textbook. In: Proceedings of the Video Track of AAAI Conference on Artificial Intelligence. AAAI (2012),
   http://www.aaaivideos.org/2012/inquire_intelligent_textbook/
8. Hedman, S.: A First Course in Logic: An Introduction to Model Theory, Proof Theory, Computability, and Complexity. Oxford Texts in Logic (2004)
9. Clark, P., Porter, B.: Building Concept Representations from Reusable Components. In: Proceedings of AAAI. AAAI Press (1997)
10. Chaudhri, V.K., Tran, S.C.: Specifying and Reasoning with Under-Specified Knowledge Base. In: International Conference on Knowledge Representation and Reasoning (2012)
11. Brachman, R.J., Levesque, H.J.: The Tractability of Subsumption in Frame-Based Description Languages. In: AAAI. AAAI Press (1984)
12. Haarslev, V., Möller, R.: The Revival of Structural Subsumption in Tableau-Based Description Logic Reasoners. In: International Workshop on Description Logics, DL 2008 (2008)
13. Baader, F.: Restricted Role-value-maps in a Description Logic with Existential Restrictions and Terminological Cycles. In: International Workshop on Description Logics, DL 2003 (2003)
14. Chein, M., Mugnier, M., Simonet, G.: Nested Graphs: A Graph-based Knowledge Representation Model with FOL Semantics. In: International Conference on Knowledge Representation, KR 1998. Morgan Kaufmann (1998)

15. Vardi, M.Y.: Why is Modal Logic so Robustly Decidable? In: Descriptive Complexity and Finite Models. DIMACS Series in Discrete Mathematics and Theoretical Computer Science, vol. 31, pp. 149–184 (1996)
16. Motik, B., Cuenca Grau, B., Horrocks, I., Sattler, U.: Representing Ontologies Using Description Logics, Description Graphs, and Rules. Artif. Intell. 173, 1275–1309 (2009)
17. Grosof, B.: The SILK Project: Semantic Inferencing on Large Knowledge (2012), http://silk.semwebcentral.org/
18. Chaudhri, V.K., Heymans, S., Wessel, M., Son, T.C.: Query Answering in Object Oriented Knowledge Bases in Logic Programming: Description and Challenge for ASP. In: Proc. of 6th Workshop on Answer Set Programming and Other Computing Paradigms (ASPOCP), Istanbul, Turkey (2013)
19. Chaudhri, V.K., Wessel, M.A., Heymans, S.: KB_Bio_101: A Challenge for TPTP First-Order Reasoners. In: KInAR - Knowledge Intensive Automated Reasoning Workshop at CADE-24, the 24th International Conference on Automated Deduction, Lake Placid, New York, USA (2013)
20. W3C OWL Working Group: OWL 2 Web Ontology Language: Document Overview. W3C Recommendation (October 27, 2009), http://www.w3.org/TR/owl2-overview/
21. Horrocks, I., Kutz, O., Sattler, U.: The Even More Irresistible $\mathcal{SROIQ}$. In: International Conference on Knowledge Representation and Reasoning (2006)
22. Chaudhri, V.K., Wessel, M.A., Heymans, S.: KB_Bio_101: A Challenge for OWL Reasoners. In: Proc. of 2nd OWL Reasoner Evaluation Workshop (ORE 2013), Ulm, Germany (2013)
23. Chaudhri, V.K., Heymans, S., Wessel, M.A.: The Bio_KB_101 Download Page (2012), http://www.ai.sri.com/halo/halobook2010/exported-kb/biokb.html

Author Index